Optimization Using Evolutionary Algorithms and Metaheuristics

Science, Technology, and Management Series

Series Editor, J. Paulo Davim, Professor, Department of Mechanical Engineering, University of Aveiro, Portugal

This book series focuses on special volumes from conferences, workshops, and symposiums, as well as volumes on topics of current interested in all aspects of science, technology, and management. The series will discuss topics such as mathematics, chemistry, physics, materials science, nanosciences, sustainability science, computational sciences, mechanical engineering, industrial engineering, manufacturing engineering, mechatronics engineering, electrical engineering, systems engineering, biomedical engineering, management sciences, economical science, human resource management, social sciences, engineering education, etc. The books will present principles, models techniques, methodologies, and applications of science, technology and management.

Advanced Mathematical Techniques in Engineering Sciences
Edited by Mangey Ram and J. Paulo Davim

Soft Computing Techniques for Engineering Optimization
Authored by Kaushik Kumar, Supriyo Roy, J. Paulo Davim

Handbook of IOT and Big Data
Edited by Vijender Kumar Solanki, Vicente García Díaz, J. Paulo Davim

For more information on this series, please visit: www.crcpress.com/Science-Technology-and-Management/book-series/CRCSCITECMAN

Optimization Using Evolutionary Algorithms and Metaheuristics

Applications in Engineering

Edited by
Kaushik Kumar
Associate Professor, Department of Mechanical Engineering
Birla Institute of Technology

J. Paulo Davim
Professor, Department of Mechanical Engineering
University of Aveiro

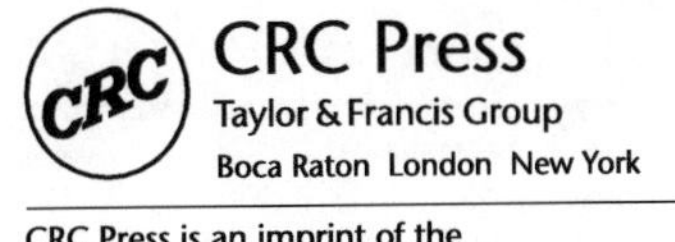

CRC Press is an imprint of the
Taylor & Francis Group, an informa business

CRC Press
Taylor & Francis Group
6000 Broken Sound Parkway NW, Suite 300
Boca Raton, FL 33487–2742

First issued in paperback 2021

ISBN 13: 978-0-367-77931-3 (pbk)
ISBN 13: 978-0-367-26044-6 (hbk)

Library of Congress Cataloging-in-Publication Data

Names: Kumar, K. (Kaushik), 1968– editor. | Davim, J. Paulo, editor.
Title: Optimization using evolutionary algorithms and metaheuristics : applications in engineering / edited by Kaushik Kumar and J. Paulo Davim.
Description: Boca Raton : Taylor & Francis, a CRC title, part of the Taylor & Francis imprint, a member of the Taylor & Francis Group, the academic division of T&F Informa, plc, 2019. | Series: Science, technology, and management series | Includes bibliographical references.
Identifiers: LCCN 2019019067 | ISBN 9780367260446 (hardback : alk. paper) | ISBN 9780429293030 (e-book)
Subjects: LCSH: Engineering economy. | Mathematical optimization. | Genetic algorithms. | metaheuristics.
Classification: LCC TA177.4 .O68 2019 | DDC 620.001/5196—dc23
LC record available at https://lccn.loc.gov/2019019067

Visit the Taylor & Francis Web site at
www.taylorandfrancis.com

and the CRC Press Web site at
www.crcpress.com

Contents

Preface......vii
Editor Biography......xi

Section I State of the Art

1 **Some Metaheuristic Optimization Schemes in Design Engineering Applications**......3
J. Srinivas

Section II Application to Design and Manufacturing

2 **AGV Routing via Ant Colony Optimization Using C#**......23
Şahin Inanç and Arzu Eren Şenaras

3 **Data Envelopment Analysis: Applications to the Manufacturing Sector**......33
Preeti and Supriyo Roy

4 **Optimization of Process Parameters for Electrical Discharge Machining of Al7075-B4C and TiC Hybrid Composite Using ELECTRE Method**......57
M. K. Pradhan and Akash Dehari

5 **Selection of Laser Micro-drilling Process Parameters Using Novel Bat Algorithm and Bird Swarm Algorithm**......83
Bappa Acherjee, Debanjan Maity, Deval Karia and Arunanshu S. Kuar

Section III Application to Energy Systems

6 **Energy Demand Management of a Residential Community through Velocity-Based Artificial Colony Bee Algorithm**......103
Sweta Singh and Neeraj Kanwar

7 **Adaptive Neuro-fuzzy Inference System (ANFIS) Modelling in Energy System and Water Resources** 117
P.A. Adedeji, S.O. Masebinu, S.A. Akinlabi and N. Madushele

Index 135

Preface

The ever-increasing demand on engineers to lower production costs to withstand competition has prompted engineers to look for rigorous methods of decision-making, such as optimization methods, to design and produce products both economically and efficiently. The term *optimize* is defined as "to make perfect." The word *optimus*—the best—was derived from Opis, a Roman goddess of abundance and fertility, who is said to have been the wife of Saturn. By her the gods designed the earth, because the earth distributes all goods (riches, goods, abundance, gifts, munificence, plenty, etc.).

Let us consider the design of a steel structure where some of the members are described by ten design variables. Each design variable represents a number of a Universal Beams (UB) section from a catalogue of ten available sections. Assuming one full structural analysis of each design takes 1 second on a computer, how much time would it take to check all the combinations of cross-sections in order to guarantee the optimum solution? The result is 317 years (10^{10} seconds). So optimization is required.

Optimization can be used in any field, as it involves formulating a process or products in various forms. It is the process of finding the best way of using the existing resources while taking into account all the factors that influence decisions in any experiment. The final product not only meets the requirements from the availability but also from the practical mass production criteria.

In computer science and mathematical optimization, a metaheuristic is a higher-level procedure or heuristic designed to find, generate or select a heuristic (partial search algorithm) that may provide a sufficiently good solution to an optimization problem, especially with incomplete or imperfect information or a limited computation capacity. Metaheuristics sample a set of solutions which is too large to be completely sampled. Metaheuristics may make few assumptions about the optimization problem being solved, and so they may be usable for a variety of problems.

Compared to optimization algorithms and iterative methods, metaheuristics do not guarantee that a globally optimal solution can be found on some class of problems. Many metaheuristics implement some form of stochastic optimization, so that the solution found is dependent on the set of random variables generated. In combinatorial optimization, by searching across a large set of feasible solutions, metaheuristics can often find good solutions with less computational effort than optimization algorithms, iterative methods or simple heuristics. As such, they are useful approaches for optimization problems.

The requirement of this book is mainly aimed at two major objectives. First, it has chapters by eminent researchers in the field providing readers about the current status of the subject. Second, algorithm-based optimization or advanced optimization techniques are mostly applied to non-engineering problems. This would serve as an excellent guideline for people in this field.

The chapters in the book have been provided by researchers and academicians who have gained considerable success in the field. The chapters of the book are segregated in three parts, namely Part I: State of the Art; Part II: Application to Design and Manufacturing; and Part III: Application to Energy Systems.

Part I contains Chapter 1; Part II contains Chapters 2 to 5; and Part III contains Chapters 6 and 7.

Part I of the book begins with Chapter 1, describing new metaheuristic optimization schemes for design engineering. It starts with the assumption that multi-constraint and multi-objective optimal problems are cumbersome to handle with available solution techniques, and metaheuristic methods provide help in such situations as they obtain an approximate solution from polynomial time algorithms. The chapter then concentrates on three such techniques applied to some engineering design problems.

Part II is dedicated to "Application to Design and Manufacturing" and contains four chapters. Chapter 2 explains Ant Colony Optimization. The technique has been used for Automated Guided Vehicle (AGV) routing. The algorithm has been utilized using C#. The work is commendable, as the black box of any software has not been used, rather the codes were generated using C#, which makes the work more versatile and robust.

Chapter 3 utilizes Data Envelopment Analysis (DEA), a mathematical optimization technique towards the manufacturing sector. The technique is being applied for selecting the best plant layout, evaluating the most effective Flexible Manufacturing System (FMS) and Advanced Manufacturing Techniques (AMT), identifying the most sustainable manufacturing system, finding the energy efficient manufacturing unit, investigating the profitability and marketability of manufacturing organizations. Hence, the chapter reviews the diverse application of DEA in the manufacturing sector and thoroughly scrutinized to state the future directions of research.

In Chapter 4, Response Surface Methodology coupled with the ELECTRE method has been proposed for optimization of the input process parameters of Electrical Discharge Machining, a non-traditional machining technique, for machining of AL7075, B4C and TiC Hybrid Composite. The significant input parameters such as pulse current (Ip), pulse duration (Ton), duty cycle (Tau) and the discharge voltage (V) are considered, and MRR, surface roughness, radial over-cut and TWR have been considered as responses for in this chapter.

In Chapter 5, the last chapter of the section, two novel bio-inspired metaheuristic algorithms are employed to investigate the laser micro-drilling process. The techniques used are very new and are the novel bat algorithm and the bird swarm algorithm. The novel bat algorithm is inspired by habitat selection behaviours of bats and their self-adaptive compensation for the

Doppler effect in echoes. The bird swarm algorithm is inspired by the social behaviours and social interactions in bird swarms. In this chapter the objective functions are developed using the response surface method (RSM). Both the algorithms are compared for their accuracy, repeatability, convergence rate and computational time and were found to be capable of predicting accurate trends of the parametric effects.

From here the book continues with Part III, grouping contributions for energy systems. Chapter 6 illustrates the application of Demand Side Management (DSM) for improvement in energy efficiency of the power grid and maintain its reliability. The load curve has been reshaped with DSM techniques and also the total cost of electricity consumption was reduced. The chapter presents a proposal for minimization of electrical cost of a residential community. Three cases have been reported for the purpose of comparison: normal load without DSM strategy in place, with DSM strategy and DSM strategy with solar PV generation. The observations reveal that it not possible to reduce the power consumption, however it is always feasible to shift the loads to cheaper pricing hours. The DSM strategy has been implemented using Velocity-based Artificial Bee Colony (VABC) algorithm and effectiveness of the load management strategy is truly reflected in the results.

The last chapter of the section and the book, Chapter 7, provides the reader with another metaheuristic technique known as the Adaptive Neuro-fuzzy Inference System (ANFIS). Its incredible ability to generalize complex nonlinear systems has been instrumental in its popularity in last two decades. The chapter dedicates for lesser-known facts about the technique (i.e. its strengths and weakness). The strengths and weaknesses of ANFIS were discussed with applications of ANFIS modelling in energy system and water resources optimization. The chapter elaborates on the black box nature of the Artificial Neural Network (ANN) with its associated pitfalls, a component of ANFIS and basic principles of the Fuzzy Inference System (FIS). Recommendations in terms of ANFIS architecture and model parameter selection were made for energy and water resources engineers on the use of ANFIS modelling technique.

First and foremost, we would like to thank God. In the process of putting this book together, the true gift of writing was very much realized and appreciated. You have given the power to believe in passion, hard work and pursue dreams. This could never have been done without faith in You, the Almighty. We would like to thank all the contributing authors without whom this would have been impossible. We would also like to thank them for believing in us. We would like to thank all of our colleagues and friends in different parts of the world for sharing ideas in shaping our thoughts. Our efforts will come to a level of satisfaction if the professionals concerned with all the fields related to optimization are benefitted. We owe a huge thanks to all our technical reviewers, editorial advisory board members, book development editor, and the team of CRC Press (A Taylor & Francis Company) for their availability to work on this huge project. All of their efforts helped to make this book complete, and we would have failed if all of them hadn't have actively supported and cooperated. Thanks to one and all.

Throughout the process of editing this book, many individuals, from different walks of life, have taken time out to help us out. Last, but definitely not least, we would like to thank them all, our well-wishers, for providing us encouragement. We would have probably given up without their support.

Kaushik Kumar
J. Paulo Davim

Editor Biography

Kaushik Kumar, B.Tech (Mechanical Engineering, REC (Now NIT), Warangal), MBA (Marketing, IGNOU) and Ph.D. (Engineering, Jadavpur University), is presently an Associate Professor in the Department of Mechanical Engineering, Birla Institute of Technology, Mesra, Ranchi, India. He has 16 years of teaching and research and over 11 years of industrial experience in a manufacturing unit of global repute. His areas of teaching and research interest are optimization, conventional and non-conventional manufacturing, CAD/CAM, rapid prototyping, quality management systems and composites. He has nine patents, 28 books, 15 edited books, 43 book chapters, 136 international journal publications, and 21 international and eight national conference publications to his credit. He is guest editor for many peer-reviewed journals and editorial boards and is a review panel member of many international and national journals of global repute. He has been felicitated with many awards and honours.

J. Paulo Davim received his Ph.D. degree in Mechanical Engineering in 1997, M.Sc. degree in Mechanical Engineering (materials and manufacturing processes) in 1991, Mechanical Engineering degree (5 years) in 1986, from the University of Porto (FEUP), the Aggregate title (Full Habilitation) from the University of Coimbra in 2005 and the D.Sc. from London Metropolitan University in 2013. He is Senior Chartered Engineer by the Portuguese Institution of Engineers with an MBA and Specialist title in Engineering and Industrial Management. He is also Eur Ing by FEANI-Brussels and Fellow (FIET) by IET-London. Currently, he is Professor at the Department of Mechanical Engineering of the University of Aveiro, Portugal. He has more than 30 years of teaching and research experience in Manufacturing, Materials, Mechanical and Industrial Engineering, with special emphasis in Machining & Tribology. He also has interest in Management, Engineering Education and Higher Education for Sustainability. He has guided large numbers of postdoc, Ph.D. and master's students as well as has coordinated and participated in several financed research projects. He has received several scientific awards. He has worked as evaluator of projects for ERC European Research Council and other international research agencies as well as examiner of Ph.D. thesis for many universities in different countries. He is the Editor in Chief of several international journals, Guest Editor of journals, books Editor, book Series Editor and Scientific Advisory for many international journals and conferences. Presently, he is an Editorial Board member of 30 international journals and acts as reviewer for more than 100 prestigious Web of Science journals. In addition, he has also published as editor (and co-editor) more than 100 books and as author (and co-author) more than 10 books, 80 book chapters and 400 articles in journals and conferences (more than 250 articles in journals indexed in Web of Science core collection/h-index 50 + /7500 + citations, SCOPUS/h-index 56 + /10500 + citations, Google Scholar/h-index 71 + /16500 +).

Section I

State of the Art

1

Some Metaheuristic Optimization Schemes in Design Engineering Applications

J. Srinivas

Associate Professor, Department of Mechanical Engineering, NIT Rourkela, Rourkela 769008, India. Email: srin07@yahoo.co.in, Ph: +91 661 2462503.

CONTENTS

1.1 Introduction 3
1.2 Optimization Schemes 5
  1.2.1 Standard Particle Swarm Optimization (PSO) 5
  1.2.2 Modified PSO 6
  1.2.3 Cuckoo Search Optimization 6
  1.2.4 Firefly Optimization 9
1.3 Case Studies in Dynamics and Design 10
  1.3.1 Bearing Parameter Identification 10
  1.3.2 Material Modelling in Nanocomposites 13
  1.3.3 Optimum Design of a Compression-Coil Spring (Kim et al. 2009) 18
1.4 Conclusions 19
References 19

1.1 Introduction

Metaheuristic is a procedure designed to generate an approach that may provide a good solution to an optimization problem, especially with incomplete available data using limited computational resources (Bianchi et al. 2009). Often, metaheuristic schemes implement some form of stochastic optimization. In combinatorial optimization process (finding optimum from a finite set), metaheuristics often predict good solutions with limited computational effort in comparison with the optimization algorithms and iterative methods, or simple heuristics. Metaheuristics are implemented in several applications including job shop scheduling, job selection, travelling salesman problems and so forth. The metaheuristic algorithms rely on exploration and exploitation. Exploration refers to the ability to diversely search in the space, while exploitation is the local search ability. There are two types of metaheuristic algorithms: (i) nature-inspired (examples like bio-inspired algorithms and physics/chemistry-based algorithms) and (ii) non-nature-inspired algorithms. Bio-inspired algorithms are based on the biological science. Two famous bio-inspired algorithms are swarm intelligence (SI) and evolutionary algorithms. Commonly employed metaheuristics for combinatorial problems are simulated annealing (Kirkpatrick et al. 1983), genetic algorithms (Holland 1973), differential evolution (Storn and Price 1997), ant colony algorithm (Dorigo and Di Caro 1999), bat search algorithm (Yang 2010), scatter search (Martí et al. 2006), tabu search (Glover and Laguna 1997), gravitational search (Rashedi et al. 2009) and so forth. Even though these approaches are very smooth and result in global optimal solutions, often they require proper selection of a large number of control parameters. For example, in genetic algorithms, population size (the number of random points), crossover rate and mutation rate are to be properly selected with several trails.

Being an end user of these algorithms, a designer may not fully focus on such parameter selection but rather pay attention on the formulation of objectives and constraints. In this regard, recently, new metaheuristic schemes have been developed such as particle swarm optimization (PSO) (Kennedy and Eberhart 1995), cuckoo search (Yang and Deb 2009), firefly algorithm (Yang 2008), artificial bee colony (Karaboga 2005), harmony search (Geem et al. 2001), spiral optimization (Tamura and Yasuda 2011) and so forth. Compared to other techniques, PSO offers several benefits. It is developed based on ideas like cognition and social behavior of particles searching for food. Each particle is associated with a velocity which helps to update the positions from time to time. Another interesting nature-inspired algorithm is the artificial bee colony (ABC) algorithm. It takes inspiration from the intelligent foraging behaviour and information-sharing capability of honeybees. In recent times, a new intelligent optimization algorithm known as cuckoo search (CS) inspired from the behaviour of cuckoos' brood parasitism and Lévy flight has been employed in many areas. It also has simple structure and is easy to implement. Another new swarm intelligence algorithm is firefly algorithm (FA). It is based on the social behaviour of flashing fireflies. In this approach, fireflies move to new positions because of the attractions among mating partners. The search continues till a certain error criterion is satisfied.

This chapter briefly presents a few mathematical optimization algorithms, and a few common case studies from dynamics and material modelling are illustrated. Some applications considered are system parameter identification from experimental data, optimal material modelling in composites and dimensional optimization of compression coil springs. Future directions of these techniques are given in conclusions.

1.2 Optimization Schemes

This section briefly outlines the mathematical basis behind some commonly used metaheuristics.

1.2.1 Standard Particle Swarm Optimization (PSO)

Particle swarm optimization is a population-based stochastic optimization approach. PSO employs two populations: (i) a population of the particles' current positions (i.e. p_i or *pbest*) and (ii) a population of the particles' best positions (i.e. p_g or *gbest*) achieved so far. The first one, particle best refers to the candidate solutions in the search space.

Here, each particle has a velocity vector V and position vector X and it moves in the search space with a velocity adjusted dynamically. The basic update equations of the particle are given as:

$$\begin{aligned} V_{id}(t+1) &= wV_{id}(t)+c_1r_1(p_{id}-x_{id})+c_2r_2(p_{gd}-x_{id}) \\ X_{id}(t+1) &= X_{id}(t)+V_{id}(t+1) \end{aligned} \tag{1.1}$$

Here, c_1 and c_2 are acceleration coefficients representing the stochastic acceleration terms that pull each particle towards *pbest* and *gbest* positions, respectively. Also, r_1 and r_2 are two uniformly distributed random numbers in the range (0, 1). The inertial weight used for balancing the global and local search is represented as w. A large value of w facilitates in global exploration, while smaller values of w leads to local exploration. Generally, w is calculated in every iteration according to the formula: $w = w_{min} + (w_{max} - w_{min})(t_{max} - t)/t$, with w_{max} and w_{min} as 0.9 and 0.4 and t_{max} is maximum generations or cycles for repeating the velocity updates. The position and velocity of ith particle is described by D-dimensional vector denoted as $X_i = [x_{i1}, x_{i2}, \ldots x_{iD}]$ for all $x_{ij} \in [x_{min}, x_{max}]$ and $V_i = [v_{i1}, v_{i2}, \ldots v_{iD}]$ for all $v_{ij} \in [v_{min}, v_{max}]$. The best previous position of the ith particle is recorded as *pbest* denoted as P_i, while the global best position of the whole swarm achieved so far is recorded as *gbest*, denoted as P_g. The algorithm is given in the following steps:

```
Initialize particle swarm randomly
While (stopping criteria is not met)
   Evaluate fitness of all particles
   for n=1 to number of particles
      find pbest, gbest
      for d=1 to number of dimensions of particle
         update the velocity of particles
         update the position of particles
      end
      end
```

```
        update the inertia weight by some corresponding strategy
        go to next generation
continue
```

Particle swarm optimization works well in terms of solution accuracy and computational speed. Also, it needs limited number of initial parameters.

1.2.2 Modified PSO

Uniformly distributed initial particles in the search space play a critical role in particle swarm optimization. Various approaches like providing chaos to the population, modifying the velocities with diversification and so forth are often used in PSO modification. To sustain the diversity of the particle swarm, the evolutionary operators like selection, crossover and mutation are used. One such approach with a mutation operator is given here. Every particle has a mutation chance which is controlled by mutation probability $p_m \in [0,1]$. For each particle, a random number between 0 and 1 is generated and the mutation is allowed on that position of the particle if this random number is less than or equal to p_m. The revised equations of modified particle position are given by:

$$\begin{aligned} \bar{X}_i(t) &= X_i(t) + \gamma * (P_{\max} - X_i(t)) \;\; if\; \gamma > 0 \\ &= X_i(t) + \gamma * (X_i(t) - P_{\min}) \;\; if\; \gamma \le 0 \end{aligned} \quad (1.2)$$

where γ is the Morlet wavelet mutation function, defined as:

$$\gamma = \frac{1}{\sqrt{a}} \exp\left(-0.5\left(\frac{\varphi}{a}\right)^2\right) \cos\left(\frac{5\varphi}{a}\right) \quad (1.3)$$

Here, ϕ is randomly generated number from [–2.5*a*, 2.5*a*] and *a* denotes the dilation parameter which is usually set to vary with the iteration of particles. There are many other variant PSO techniques available in literature for improving the performance of PSO.

1.2.3 Cuckoo Search Optimization

Another interesting algorithm used in recent times is cuckoo search (CS) optimization, which is a nature-inspired metaheuristic algorithm, based on the brood parasitism of some cuckoo species, along with Lévy flight random walks. Cuckoos are unique species because of their special reproduction strategy. Cuckoos always lay eggs in host nests. To increase the hatching probability, some of the host bird's eggs will be pushed out of the nest. If a cuckoo's egg is found by the host bird, it will abandon this nest and build a new nest at other place. In the CS algorithm, each egg in a nest represents a solution and a cuckoo egg represents a new solution. The aim is to use the new and potential solutions (cuckoos) to replace worst solutions in the nests. The CS algorithm is focused on three idealized rules (Yang and Deb 2009):

1. Each cuckoo lays one egg at a time and dumps it in a randomly chosen nest.
2. The best nest with high quality of eggs (solutions) will carry over to the next generations.
3. There are fixed number of available host nests and a host bird can discover an alien egg with a probability $p_a \in [0, 1]$. In this case, the host bird can either throw the egg away or abandon the nest and construct a completely new nest.

For simplicity, the last assumption can be simulated by the fraction of the population size of worst nests that are replaced by new random nests. To start with the algorithm, the initial positions of nests are predicted by a set of randomly assigned values to control variables:

$$X^t = X_{min} + rand * (X_{max} - X_{min}) \tag{1.4}$$

where X^t denotes the initial vector for the ith nest and X_{max} and X_{min} are the upper and lower bounds for the variables under consideration. In the next step, the entire nests are replaced by new cuckoo eggs except the best one. Cuckoo moves from the current nest to the new one using a random step length which is drawn from a Lévy distribution. The new nest position is determined using Lévy flights based on their quality according to

$$X^{t+1} = X^t + \alpha.S.\left(X^t - X^t_{best}\right).r \tag{1.5}$$

where S is step size; α ($\alpha > 0$) represents a step size scaling factor; r is random number from standard normal distribution and X^t_{best} is the location of best nest. S is calculated by random walk using Lévy flights and Mantegna's algorithm as:

$$S = \frac{u}{|v|^{1/\beta}} \tag{1.6}$$

where u and v have a normal distribution with zero means and associated variance, as given by the following equation:

$$\sigma_u = \left(\frac{\Gamma(1+\beta)\sin(\pi\beta/2)}{\Gamma(1+\beta/2)\times\beta\times 2^{(\beta-1)/2}}\right)^{1/\beta}, \sigma_v = 1 \tag{1.7}$$

Here, β is a Lévy parameter (Lévy flights exponent) selected in the range (Bianchi et al. 2009; Kirkpatrick et al. 1983), and considered as 3/2 in the present work. Also, Γ is the standard gamma function.

The second rule passes the best solutions to the next generation. Finally, the last rule carries out global search and can be seen as the mutation operator, where the worst solutions are replaced with newly generated solutions. It prevents in trapping at a local minimum. To generate solutions with simple random walk, the following update rule is used.

$$X_i^{t+1} = X_i^t + r \otimes H(p_a - \varepsilon) \otimes (X_j^t - X_k^t) \tag{1.8}$$

where $\otimes$ is an entry-wise multiplication, p_a is the switching probability, which is responsible for balancing between local and global optimization, $H(.)$ is the Heaviside function, r and ε are two random numbers with uniform distribution and X_j^t, X_k^t are two solutions randomly selected. This complete process is repeated to get a best solution. After each iteration the input data is updated with new population and the objective function values are obtained. The algorithm is summarized as given below:

```
begin
   —Objective function f(X)
   —Generate initial population of n host nests Xi
(i = 1,2, . . . n)
while (t<MaxGeneration) or (stop criterion)
   —Get a cuckoo randomly by Levy flights
   —Evaluate its quality or fitness Fi
   —Choose a nest among n (say j) randomly
if (Fi>Fj)
Replace j by new solution;
end
   —A fraction (pa) of worst nests are abandoned and new ones
are built;
   —Keep the best solutions (or nests with quality solutions)
   —Rank the solutions and find the current global best
end while
   —Post process results and visualization
end
```

By providing more diversity change than this standard cuckoo search, the global convergence has improved considerably which also prevents in trapping at local optimal solutions. Chaos-enhanced cuckoo search optimization algorithm is one kind where chaos is embedded into the standard cuckoo search algorithm at the initialized host nest location and Lévy flight parameters.

1.2.4 Firefly Optimization

The firefly algorithm (FA) was inspired by the flashing patterns of fireflies (Fister et al. 2018). Each firefly in the population represents a candidate solution in the *D*-dimensional search space. Due to the attractions among fireflies, they can move towards the other better positions and find better candidate solutions. In FA, the attraction is determined by the brightness, which is taken as the value of the objective function. The attractiveness β between two fireflies is relative to their distance *r*. As the distance increases, the attractiveness gradually decreases. Let X_i be the *i*th firefly in the population, where $i = 1, 2, ..., ps$ with *ps* as the population size. The attractiveness β between two fireflies X_i and X_j can be calculated as: $\beta = \beta_0 \exp(-\gamma r_{ij}^2)$, where $r_{ij} = \|X_i - X_j\|$ is the second order norm. The parameter β_0 is the attractiveness at the distance $r = 0$, and γ is the light absorption coefficient. Often, γ is set as $1/\Gamma^2$, where Γ is the length scale for designed variables. For each firefly X_i, a comparison is made with other all fireflies X_j, where $j = 1, 2, ..., ps$. If X_j is brighter (better) than X_i, then X_i moves towards X_j by the attraction. The movement or update of X_i is given as

$$X_i(t+1) = X_i(t) + \beta_0 e^{-\gamma r_0^2}\left(X_j(t) - X_i(t)\right) + \alpha\varepsilon \tag{1.9}$$

Here $\alpha \in [0,1]$ is the step factor, while ε is a random value within [−0.5, 0.5].

Generally, the maximum number of generations G_{max} is taken as stopping criterion. For minimization of function *f*(X), following steps are followed:

```
Randomly initialize the population of ps fireflies (solutions) Xi(t)
Compute the fitness values of each firefly

While t ≤ Gmax do
   for i = 1 to ps do
      for j = 1 to ps do
         if f(Xj) < f(Xi) then
      move Xi towards Xj as per above equation
       calculate the fitness value of Xi (t+1)
      end
   end
   end
  increase t to t+1
end
```

The algorithm requires high computational time. Often, through the attraction, fireflies gradually approach to the converged states in a number of generations. Then the distance between fireflies gradually decreases to zero. To avoid this situation, following changes are made:

the parameter $\beta_0(t + 1) = r_1$ (a uniformly distributed random number) if $r_2 < 0.5$ (where r_2 is another random number);

otherwise $\beta_0(t + 1) = \beta_0(t)$.

Similarly, $\alpha(t+1) = \alpha(t)(1 - t/G_{max})$ allows faster convergence to the algorithm. Commonly, initial values considered are $\beta_0(0) = 1$ and $\alpha_0(0) = 0.5$.

1.3 Case Studies in Dynamics and Design

This section presents three different application areas for optimization problems in design, materials engineering and dynamics.

1.3.1 Bearing Parameter Identification

In dynamics of rotating system, often the bearing forces are important and in practice they are highly nonlinear functions of bearing displacements. Complexity of the analysis can be minimized by computing equivalent linear bearing force (stiffness and damping) coefficients.

Usually, experimental response data in a rotor is measured at the bearings using accelerometers. As these vibration signals are arrived from the practical system, they are realized to be derived from the system undergoing nonlinear bearing forces. Thus, if an equivalent linear system model response is fitted over the experimental response, it is possible to derive the correct set of unknown linear force coefficients. That means an error function formulation and its minimization is the task in this problem.

In the present case, experimental response data is captured from a rotating dual disk rotor shaft supported on two similar oil-film bearings. By using an equivalent linear physical system representation, the bearing forces are expressed in terms of displacements and velocities at bearing coordinates as:

$$\begin{Bmatrix} \tilde{F}_{Bx} \\ \tilde{F}_{By} \end{Bmatrix} = \begin{bmatrix} c_{xx} & c_{xy} \\ c_{yx} & c_{yy} \end{bmatrix} \begin{Bmatrix} \dot{x}_B \\ \dot{y}_B \end{Bmatrix} + \begin{bmatrix} k_{xx} & k_{xy} \\ k_{yx} & k_{yy} \end{bmatrix} \begin{Bmatrix} x_B \\ y_B \end{Bmatrix} \tag{1.10}$$

where c_{xx}, c_{yy}, k_{xx}, k_{yy} and c_{xy}, c_{yx}, k_{xy}, k_{yx} terms respectively represent the unknown direct and cross-coupled bearing force coefficients, the suffix B denotes the bearing support location. By substituting these bearing forces in the rotor dynamic model $[M]\{\ddot{Z}\}+[C]\{\dot{Z}\}+[K]\{Z\}=\{F_B\}+\{F\}$ and converting into frequency domain, these system of equations can be written as (Kim et al. 2007)

$$\begin{Bmatrix} F_x \\ F_y \end{Bmatrix} - \omega^2[\tilde{\mathbf{M}}] \begin{Bmatrix} X \\ Y \end{Bmatrix} = \begin{bmatrix} H_{xx} & H_{xy} \\ H_{yx} & H_{yy} \end{bmatrix} \begin{Bmatrix} X \\ Y \end{Bmatrix} \tag{1.11}$$

Here, $H_{ij}(\omega) = k_{ij}(\omega) + i\omega c_{ij}(\omega)$ (with $i,j = x,y$) is the impedance function, (F_x, F_y) and (X,Y) are the discrete Fourier transforms of external forces and displacements respectively. By knowing the component displacements in bending directions {Z} at bearing node, it is possible to compute the eight direct coefficients corresponding to each of the two bearings. In order to obtain these parameters, error function of amplitudes of displacements at the bearing nodes is considered as:

$$E = \sum_{n=1}^{sp} \left(X_{\exp} - X_{fe}\right)^2 + \left(Y_{\exp} - Y_{fe}\right)^2 \tag{1.12}$$

where X_{exp} and Y_{exp} are the experimentally obtained measured amplitudes at a bearing, while X_{fe} and Y_{fe} are the numerically obtained bearing amplitudes in terms of unknown equivalent linear bearing stiffness and damping parameters. Here, *sp* represents the number of samples considered in the frequency range of interest. The upper and lower bounds are considered as $k_i \in [k_{min}, k_{max}]$ and $c_i \in [c_{min}, c_{max}]$ with $I = x,y$. Figure 1.1 shows the rotor model with two discs and their geometric parameters.

Table 1.1 shows the geometric and material parameters of the rotor system used for linear bearing parameter-finite element model.

Figure 1.2 shows the time history and Fast Fourier Transform (FFT) plots obtained from the experimental analysis at the left bearing node in two directions at a rotor speed of 900 rpm.

The error function is then formulated and minimized using modified particle swarm optimization method and the stiffness and damping coefficients are estimated. Figure 1.3 shows the identified parameters at the two bearing nodes in two directions.

Using these coefficients, the FFT spectra obtained from finite element model are shown in Figure 1.4. It is seen that the first resonant peak occurs at 159 Hz, coming close to the experimentally obtained value (157.2 Hz).

In the present work, $c_1 = c_2 = 2.0$, $w_{max} = 1.4$ and $w_{min} = 0.8$ were selected. Also, the bounds for velocity V_{max} and V_{min} parameters may be set for each

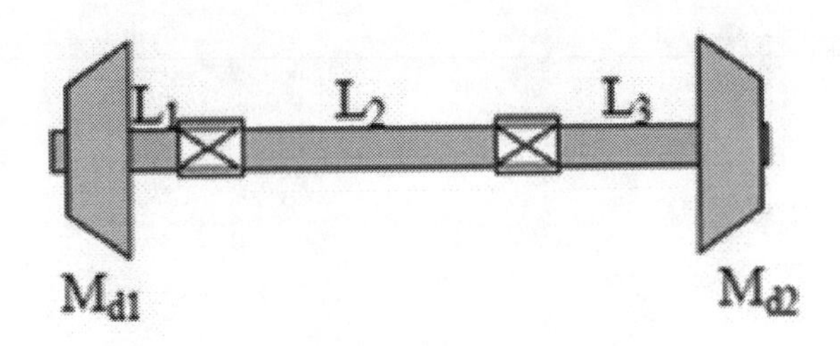

FIGURE 1.1
FCNRotor-disk model

TABLE 1.1

Material and geometric data for numerical modelling (Kim et al. 2007)

Properties	Value
Density of shaft material (kg/m^3)	7,800
Mass of Disc1, M_{D1} (kg)	1.4
Mass of Disc2, M_{D2} (kg)	1
Diameter of shaft, D_{sh} (m)	0.016
Length of the shaft (m)	0.48
Young's modulus, E (GPa)	200
Distance between the bearings (m)	0.22
Distance from Disc1 to left bearing (m)	0.09
Distance from Disc2 to right bearing (m)	0.09

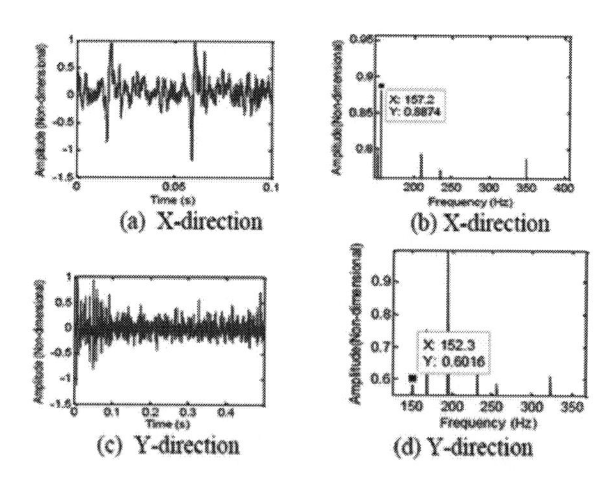

FIGURE 1.2
Time history and FFT plots at the left bearing in both X and Y directions

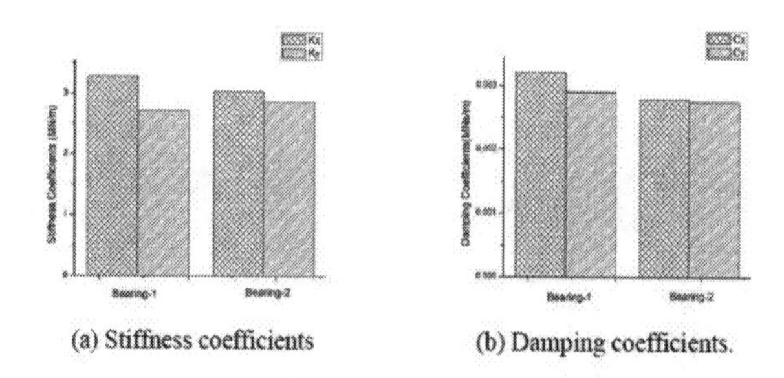

FIGURE 1.3
Identified bearing parameters

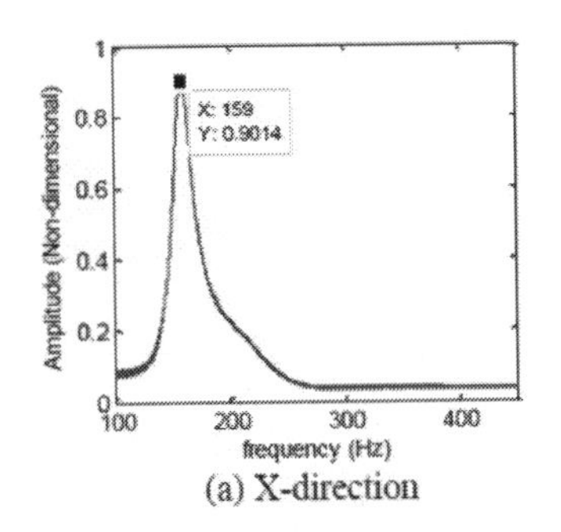

(a) X-direction

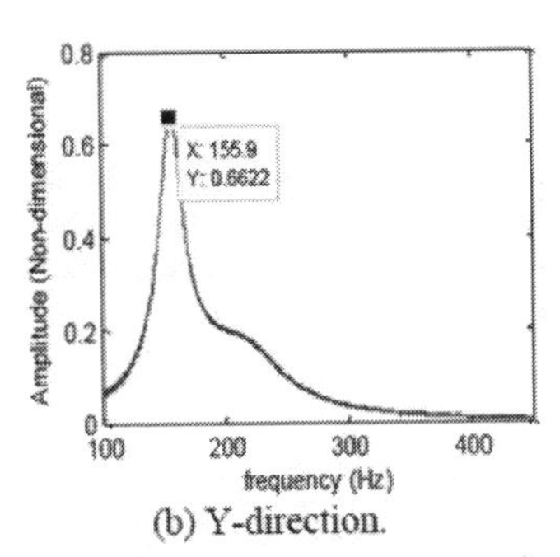

(b) Y-direction.

FIGURE 1.4
FFT plots at left bearing

particle to control the velocity within a reasonable range. In this study, V_{max} and V_{min} are set as 10% of the upper and lower values, respectively.

1.3.2 Material Modelling in Nanocomposites

Recently, nanocomposite materials are gaining importance in various fields of engineering Shojaeefard and Khalkhali (2014). There are several factors in the fabricating stage affecting the overall properties of composite. In present study, a Carbon NanoTube (CNT) reinforced polymer composite material is considered for analysis. Various configurations including variable length CNTs of different volume fractions as well as with interphase region between CNT and matrix domain are accounted to know the effects on stiffness and thermal properties of composite. Studies are conducted for single walled (SWCNTs) and multi-walled CNT (MWCNTs) reinforcement. Along with interphase, some other factors like agglomeration and waviness of CNTs also play an important role on elastic modulus and thermal conductivity of composite. Based on the analytical improved models (Halpin-Tsai and effective medium approximation model), it was observed that the volume fraction (V_{cnt}), aspect ratio (AR), interphase thickness ratio (TR), waviness ratio (w) and inclusion diameter factor (f) have more influence on the resultant elastic modulus and thermal conductivity (Puneet Kumar 2018). The elastic modulus of a composite, E_c, due to insertion of CNT (having elastic modulus E_{cnt}) is given by:

$$E_c = \eta_l \eta_o E_{cnt} V_{cnt} + E_m (1 - V_{cnt}) \tag{1.13}$$

where V_{cnt} is the volume fraction of the filler, E_m the elastic modulus of matrix, η_l is the length efficiency factor and η_o is the Krenchel orientation factor. The expression for η_l in terms of length and diameters of CNT (L_{cnt} and D_{cnt}) is given by

$$\eta_l = 1 - \frac{\tanh(a\,L_{cnt}/D_{cnt})}{a\,L_{cnt}/D_{cnt}} \tag{1.14}$$

where

$$a = \sqrt{\frac{-3E_m}{2E_{cnt}\ell n(V_{cnt})}} \tag{1.15}$$

The length efficiency parameter approaches 1 for L_f/D_f>10. For one-dimensional materials, $\eta_o = 1$ for perfectly aligned fibers or nanotubes, but 3/8 for fibers or nanotubes oriented randomly in the 2-D plane and 1/5 for material with random 3-D fibers or nanotubes. The rule of mixtures provides a simple method to estimate the effective elastic properties of the composite; however, it is just a first approximation without taking into account other factors, such as Poisson's ratio. The improvement of mechanical properties in polymer nanocomposites is attributed to strong interfacial adhesion/interaction between polymer matrix and nanoparticles, which suitably transfers stress from the continuous matrix to the nano-filler. Also, small nanoparticles and their good dispersion play positive role in behaviour of polymer nanocomposites. From a theoretical point of view, conventional models such as those of Halpin-Tsai and Guth fail to properly account these parameters and cannot give correct calculations of mechanical properties of polymer nanocomposites. An interphase forms in polymer nanocomposites due to high interfacial area and strong interfacial interactions between polymer and nanoparticles. The interphase has different properties from polymer matrix and nanoparticles phases, which significantly affects the properties of polymer nanocomposite. The volume fraction of interphase (V_i) for nanocomposites containing cylindrical nanoparticles can be calculated by:

$$V_{\text{int}} = \left[\left(\frac{R + R_{\text{int}}}{R}\right)^2 - 1\right]V_{cnt} \tag{1.16}$$

where R and R_{int} are the radius of nanoparticles and interphase thickness, respectively. If $R_{int} = 0$, $V_{int} = 0$, which indicates the absence of interphase in polymer nanocomposites. By addition of interphase effects to the Halpin-Tsai model, the effective elastic modulus can be written as:

$$\frac{E_c}{E_m} = \frac{1 + \xi\eta_f V_{cnt} + \xi\eta_{\text{int}} V_{\text{int}}}{1 - \eta_f V_{cnt} - \eta_{\text{int}} V_{\text{int}}} \tag{1.17}$$

where

$$\eta_f = \frac{(E_{cnt\backslash} / E_m - 1)}{(E_{cnt} / E_m + \xi)} \tag{1.18}$$

$$\eta_{\text{int}} = \frac{(E_{\text{int}} / E_m - 1)}{(E_{\text{int}} / E_m + \xi)} \tag{1.19}$$

Here E_{int} is Young's modulus of interphase. During processing of nanocomposites, waviness, agglomeration and random orientations as well as interphase effects are major factors influencing the overall performance of composite considerably. The modified Halpin-Tsai model for accounting the nonlinear elastic behaviour with volume fraction is represented for a randomly oriented case as

$$E_c = \left\{\frac{3}{8}\right\} E'_L + \left\{\frac{5}{8}\right\} E'_T \tag{1.20}$$

where

$$E'_L = \left[\frac{1 + \xi'_L \eta'_L V_{cnt}}{1 - \eta'_L v_{cnt}}\right] E_m \tag{1.21a}$$

$$E'_T = \left[\frac{1 + \xi'_T \eta'_T V_{cnt}}{1 - \eta'_T v_{cnt}}\right] E_m \tag{1.21b}$$

$$\eta'_L = \left\{\frac{(K_i K_w E_{cnt} / E_m) - 1}{(K_i K_w E_{cnt} / E_m) + \xi'_L}\right\} \tag{1.21c}$$

$$\eta'_T = \left\{\frac{(K_i K_w E_{cnt} / E_m) - 1}{(K_i K_w E_{cnt} / E_m) + \zeta'_T}\right\} \tag{1.21d}$$

$$\xi'_L = 2\left(\frac{L_f}{D_f}\right) K_{agg},\ \xi'_T = 2 \tag{1.21e}$$

where E_{cnt}, E_m and E_c indicate the elastic modulus of CNT, polymer matrix and composite respectively. Also, K_i, K_w, K_{agg} are known as interphase, waviness and agglomeration factors, respectively. These factors depend on the geometric and processing characteristics of CNT reinforced polymer composites and explained here.

Microscopic observations have shown that most of CNTs result into curved shapes within the polymer matrix during their processing. This is due to very low bending stiffness and high aspect ratio of CNTs. An analytical formula is developed in terms of dependent variables as:

$$K_w = \frac{dz}{ds} = \frac{1}{\sqrt{1 + 4\pi^2 w^2 \sin(2\pi L_n)}} \tag{1.22}$$

where K_w is a waviness factor, which depends on the waviness ratio of CNT ($w = A/\lambda$) and normalized length ($L_n = z/\lambda$). In present work, shape of wavy CNT is considered as *cosine* form with waviness ratio ranging from 0 to 1 and normalized length varies from 0.1 to 0.5. Uniformly dispersed CNTs within polymer matrix is essential to achieve good thermo-mechanical properties. However, it is observed that during high energy ultra-sonication and shear mixing, de-agglomeration and CNT breakages occur simultaneously. Hence, CNTs with higher aspect ratio may break into small pieces during mixing and fabrication process of polymer nanocomposite. In other words, dispersion index rises while the mean tube length reduces during the mixing process. So the sonication time and sonication energy have a considerable effect on the overall properties. A formulation of length and diameter of CNTs is developed to include the influence of sonication time on mechanical properties of the nanocomposite as follows:

$$L_f = L_{cnt} t^{-0.248} \tag{1.23a}$$

$$D_f = D_{cnt} + D_{inclusion} t^{-0.248} \tag{1.23b}$$

where L_f and D_f are the final length and diameter of CNT fibre after sonication process. D_{cnt}, L_{cnt} are initial diameter, length of CNTs and t is the sonication time required. Also, $D_{inclusion} = (f \times D_{cnt})$ represents the diameter of inclusion or CNTs bundle and f is inclusion formation factor exist at initial stage of sonication process. It is considered that initial length and diameter of agglomerates decrease during the mixing and can be represented in terms of CNT's geometry. Non-uniform dispersion and agglomeration of CNTs in polymer composites often lead to deterioration of the mechanical and thermal strengths. To ensure uniform distribution of the CNTs within the solution mixture, it is often sonicated to disperse the CNTs prior to curing of composite. Interphase is another factor of importance. In fact, its properties vary as a function of distance from surface of the CNT to polymer matrix. To determine the influence of interphase on elastic properties of CNT reinforced composite, an interphase factor K_i is defined as:

$$K_i = 1 - e^{\left\{\left(\frac{ER}{TR}\right)\log\left(\frac{1}{(L_f/D_f)}\right)\right\}} \tag{1.24}$$

where $ER = E_{\text{int}} / E_{cnt} = k_{\text{int}} / k_{cnt}$ is effective stiffness or conductivity ratio and $TR = t_{\text{int}} / t_{cnt}$ is the thickness ratio. This factor K_i depends on the properties of CNTs and surrounding polymer i.e. effective ratio (*ER*) and interphase thickness ratio (*TR*), respectively. Thickness of CNT t_{cnt} is assumed as 0.34 nm and thickness ratio varies in terms of t_{cnt}. In order to maximize the elasticity of the composite system, optimum values of the most influencing five parameters are to be selected. Both cuckoo search and particle swarm optimization methods are employed in optimization. The following parameters are considered in CS optimization: nest size = 25, $p_a = 0.25$, $\beta = 1.5$, $\alpha = 0.01$. A convergence trend of optimum elastic modulus is plotted in Figure 1.5(a) using cuckoo search algorithm for different number of nest size. A fast convergence towards optimum elastic modulus of composite can be seen for higher nest size while other parameters kept constant. It reveals that maximum value of objective function is approximately 30 cycles. Figure 1.5(b) are plotted to investigate the effectiveness and stability of the developed algorithm by changing different algorithm parameters.

Figure 1.6 shows the comparative analysis of optimizing with CS and PSO. It is observed that the CS algorithm gives comparable results in lesser time.

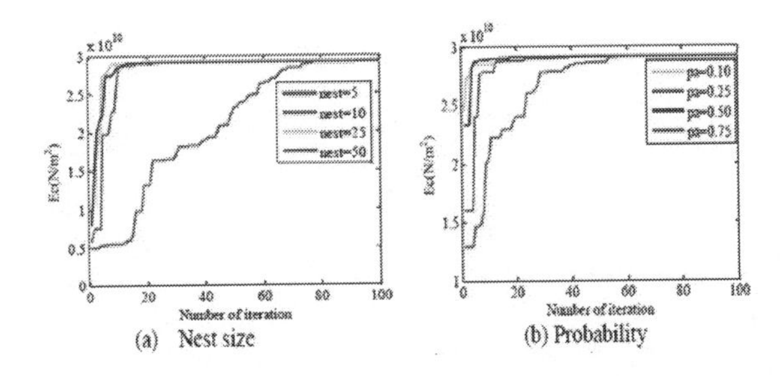

FIGURE 1.5
Convergence analysis of E_c using CS algorithm

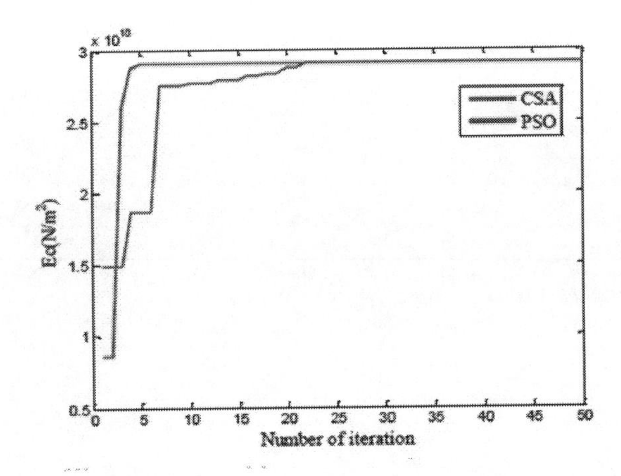

FIGURE 1.6
Convergence of elastic modulus

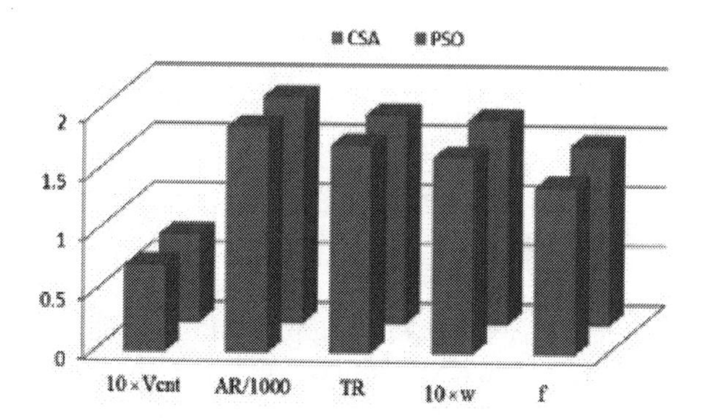

FIGURE 1.7
Optimized values of design parameters using two schemes

The corresponding optimized variables are shown in Figure 1.7 as a bar chart.

1.3.3 Optimum Design of a Compression-Coil Spring (Kim et al. 2009)

Optimal design of tension-compression coil springs is a standard problem. Here, the objective is to minimize its weight, subjected to certain maximum values of deflection, shear stress, and surge frequency limit with bounds on the outside diameter. There are three design variables: number of coils (N), winding diameter (D) and wire diameter (d). It is formulated as:

Minimize

$$f(X) = (N+2)Dd^2 \tag{1.25}$$

Subject to

$$g_1(X) = 1 - \frac{ND^3}{71785d^4} \le 0 \tag{1.26}$$

$$g_2(X) = \frac{4D^2 - Dd}{12566(Dd^3 - d^4)} + \frac{1}{5108d^3} - 1 \le 0 \tag{1.27}$$

$$g_3(X) = 1 - \frac{140.45d}{ND^2} \le 0 \tag{1.28}$$

$$g_4(X) = \frac{D+d}{1.5} - 1 \le 0 \tag{1.29}$$

$$X \in [N, D, d] \text{ with } 2 \le N \le 15, 0.25 \le D \le 1.3,\ 0.05 \le d \le 2 \tag{1.30}$$

Here, the problem is solved as equivalent unconstrained one as follows:

Minimize

$$F(X) = F(X) = \begin{cases} g_{max}(X) \text{ if } g_{max}(X) > 0 \\ a\tan(f(X)) - \dfrac{\pi}{2}, \text{otherwise} \end{cases} \tag{1.31}$$

where $g_{max}(X) = \max[g_1(X), g_2(X), g_3(X), \ldots]$.

By considering the following parameters of firefly algorithm $\gamma = 1$ and $\alpha = 0.25$, the solution obtained is $N^* = 11.16$, $D^* = 0.3589$m and $d^* = 0.0517$m with $f(X) = 0.01266$. These are close to the results obtained by the PSO approach (Kim et al. 2009).

1.4 Conclusions

This chapter has focused on the three modern metaheuristic optimization schemes, namely, particle-swarm optimization, cuckoo search optimization and firefly optimization. Some possible modifications of basic methods were explained. Three application case studies were considered from design, materials and dynamics. In the first example, the linear bearing force coefficients were identified from measured vibration responses. In next case, the required optimum process parameters during material modelling for increasing the effective elastic modulus of CNT reinforced polymer nanocomposite. In the last case, the prediction of optimal coil parameters of a compression spring under some constraints was presented. All these are very convenient tools, and user-friendly computer codes give the solutions very accurately. As a future scope of these methods, we have sorted out micro-equivalent models so as to reduce the number of particles to be considered. Also, use of hybrid techniques may be very beneficial. Constrained, multi-objective formulations are to be tested with a larger number of in-depth case studies.

References

Bianchi, L., Marco, D., Luca, M.G., Walter, J.G., & Gutjahr, J. (2009). A survey on metaheuristics for stochastic combinatorial optimization. *Natural Computing: An International Journal, 8*(2), 239–287.

Dorigo, M., & Di Caro, G. (1999). Ant colony optimization: A new meta-heuristic. *Proceedings of the 1999 Congress on Evolutionary Computation (CEC 99)*, IEEE, pp. 1470–1477.

Fister, I., Fister, I., Jr., Yang, X.S., & Brest, J. (2018). A comprehensive review of firefly algorithms. *Swarm and Evolutionary Computation*. doi:10.1016/j.swevo.2013.06.001.

Geem, Z.W., Kim, J.H., & Loganathan, G.V. (2001). A new heuristic optimization algorithm: Harmony search. *Simulation, 76*(2), 60–68.

Glover, F., & Laguna, M. (1997). *Tabu search*. Norwell, MA: Kluwer Academic Publisher.

Holland, J.H. (1975). *Adaption in natural and artificial systems*. Ann Arbor, MI: University Michigan Press.

Karaboga, D. (2005). An idea based on honey bee swarm for numerical optimization, technical report. *Technical Report-tr06*, Erciyes University, Engineering Faculty, Computer Engineering Department.

Kennedy, J., & Eberhart, R.C. (1995). Particle swarm optimization. *IEEE International Conference on Neural Networks*, 1942–1948.

Kim, T-H., Maruta, I., & Sugie, T. (2009). A simple and efficient constrained particle swarm optimization and its application to engineering design problems. *Proceedings of the Institution of Mechanical Engineers, Part C: Journal of Mechanical Engineering Science, 224*, 389–400.

Kim, Y.H., Yang, B.S., & Tan, A.C.C. (2007). Bearing parameter identification of rotor-bearing system using clustering-based hybrid evolutionary algorithm. *Structural and Multidisciplinary Optimization, 33*, 493–506.

Kirkpatrick, S., Gelatt, C.D., Jr., & Vecchi, M.P. (1983). Optimization by simulated annealing, *Science, 220*(4598), 671–680.

Kumar, P., & Srinivas, J. (2018). Elastic and thermal property studies of CNT reinforced epoxy composite with waviness, agglomeration and interphase effects. *International Journal of Materials Engineering Innovation, 9*, 158. doi:10.1504/IJMATEI.2018.10014996.

Martí, R., Laguna, M., & Glover, F. (2006). Principles of scatter search. *European Journal of Operational Research, 169*(2), 359–372.

Rashedi, E., Nezamabadi-Pour, H., & Saryazdi, S. (2009). GSA: A gravitational search algorithm. *Information Science, 179*(13), 2232–2248.

Shojaeefard, M.H., Khalkhali, A., & Khakshournia, S.H. (2014). Multi-objective optimization of a CNT/Polymer nanocomposite automotive drive shaft. *The 3rd International Conference on Design Engineering and Science, ICDES 2014*, Pilsen, Czech Republic, August 31–September 3.

Storn, R., & Price, K. (1997). Differential evolution—a simple and efficient heuristic for global optimization over continuous spaces. *Journal of Global Optimization, 11*, 341–359.

Tamura, K., & Yasuda, K. (2011). Primary study of spiral dynamics inspired optimization. *IEEJ Transactions on Electrical and Electronic Engineering, 6*(S1), 98–100.

Yang, X.S. (2008). *Nature-inspired metaheuristic algorithms*. Bristol, UK: Luniver Press, pp. 79–86.

Yang, X.S. (2010). A new metaheuristic bat-inspired algorithm. *Nature Inspired Cooperative Strategies for Optimization, 284*, 65–74.

Yang, X.S., & Deb, S. (2009). Cuckoo search via Lévy flights. *Proceedings of World Congress on Nature & Biologically Inspired Computing*, India, pp. 210–214.

Section II

Application to Design and Manufacturing

2

AGV Routing via Ant Colony Optimization Using C#

Şahin Inanç

Bursa Uludag University, Lecturer, sahininanc@uludag.edu.tr

Arzu Eren Şenaras

Bursa Uludag University, Research Assistant Dr, arzueren@uludag.edu.tr

CONTENTS

2.1 Introduction 23
2.2 A Short Literature Review 24
2.3 Ant Colony Optimization (ACO) 25
2.4 ACO Application via C# 26
2.5 Conclusion 28
References 31

2.1 Introduction

An AGV (automated guided vehicle) consists of a mobile robot used for transportation and automatic material handling, for example for finished goods, raw materials, and products in process. The design and operation of AGV systems are highly complex due to high levels of randomness and the large number of variables involved. This complexity makes simulation an extremely useful technique in modelling these systems (Negahban and Smith 2014). The AGV has the function to ensure efficient flow of materials within the production system. Production systems must be flexible and must allow the dynamic reconfiguration of the system. The AGV is a key component to achieve the objectives of an FMS. This means that the AGV should provide the required materials to the appropriate workstation, at the right time and in the right amount, otherwise the production system will not perform well, making it less efficient, generating less profit or increasing the operating costs. In an FMS system, the AGV has the following advantages (Leite et al. 2015):

- Driverless operation
- More efficient control of the production
- Diminishing of the damages caused by manual material handling.

An AGV is a driverless material handling system used for horizontal movement of materials. AGVs were introduced in 1955. The use of AGVs has grown enormously since their introduction. The number of areas of application and variation in types has increased significantly. AGVs can be used in inside and outside environments, such as manufacturing, distribution, transshipment and (external) transportation areas. At manufacturing areas, AGVs are used to transport all types of materials related to the manufacturing process (Fazlollahtabar and Saidi-Mehrabad 2015).

An AGV consists of a mobile robot used for transportation and automatic material handling, for example for finished goods, raw materials, and products in process. The design and operation of AGV systems are highly complex due to high levels of randomness and the large number of variables involved. This complexity makes simulation an extremely useful technique in modelling these systems (Negahban and Smith 2014).

2.2 A Short Literature Review

Kulatunga et al. (2006) studied a metaheuristic-based ant colony optimization (ACO) technique for simultaneous task allocation and path planning of AGV in material handling. They found that ACO solutions have slightly better performance than that of simulated annealing algorithm.

Udhayakumar and Kumanan (2010) studied to find the near optimum schedule for two AGVs based on the balanced workload and the minimum traveling time for maximum utilization.

Leite et al. (2015) investigated the utilization rate of an AGV system in an industrial environment and evaluated the advantages and disadvantages of the project. They used the simulation software Promodel 7.0 to develop a model. Their model aims to analyze and optimize the use of AGVs. Problems were identified and solutions were adopted by the authors according to the results obtained from the simulations.

Fazlollahtabar and Saidi-Mehrabad (2015) categorized the methodologies into mathematical methods (exact and heuristics), simulation studies, metaheuristic techniques and artificial intelligence–based approaches.

Wang et al. (2016) studied a scheduling problem in the FMS in which orders require the completion of different parts in various quantities. The orders arrive randomly and continuously, and they all have predetermined due dates. Two scheduling decisions were considered in this study: launching parts into the system for production and determining the order sequence for collecting the completed parts.

Hana and Gabriel (2016) aimed to present the possibilities of computer simulation methods for obtaining data for a full-scale economic analysis implementation.

Vavrika et al. (2017) studied methods for the determination of the number of automated guided vehicles and choosing the optimal internal company logistics track. The simulation results of the logistics system were various in terms of increasing the use of operation areas, optimized material supply, and a created layout that would be able to flexibly respond to future company requirements.

Demesure et al. (2017) proposed motion planning and the scheduling of AGVs in an FMS. Numerical and experimental results are provided to show the pertinence and the feasibility of the proposed strategy.

2.3 Ant Colony Optimization (ACO)

In the early 1990s, ant colony optimization (ACO) was introduced by M. Dorigo and colleagues as a novel, nature-inspired metaheuristic for the solution of hard combinatorial optimization (CO) problems. ACO belongs to the class of metaheuristics, which are approximate algorithms used to obtain good-enough solutions to hard CO problems in a reasonable amount of computation time. Other examples of metaheuristics are tabu search, simulated annealing, and evolutionary computation. The inspiring source of ACO is the foraging behaviour of real ants. When searching for food, ants initially explore the area surrounding their nest in a random manner. As soon as an ant finds a food source, it evaluates the quantity and the quality of the food and carries some of it back to the nest. During the return trip, the ant deposits a chemical pheromone trail on the ground. The quantity of pheromone deposited, which may depend on the quantity and quality of the food, will guide other ants to the food source. As it has been shown in, indirect communication between the ants via pheromone trails enables them to find the shortest paths between their nest and food sources. This characteristic of real ant colonies is exploited in artificial ant colonies in order to solve CO problems (Dorigo and Blum 2005).

Ant colony algorithms were first proposed by Dorigo and colleagues as a multi-agent approach to difficult combinatorial optimization problems such as the traveling salesman problem and the quadratic assignment problem. There is currently much ongoing activity in the scientific community to extend and apply ant-based algorithms to many different discrete optimization problems. Recent applications cover problems such as vehicle routing, job shop scheduling, quadratic assignment problems and so on (Kulatunga et al. 2006).

2.4 ACO Application via C#

This work was used to calculate the shortest path in this study by passing through all the nodes according to the ant colony algorithm and providing a closed loop by returning to the starting point, provided that the dropped node does not come back. These types of problems are referred to as traveling salesman problems.

In this study, ten nodes were selected as an application. The nodes were used to determine the route of the AGV vehicle.

In practice, a cluster structure (array structure) is created for each node to which nodes can be traversed. Routes between the nodes are defined as an array and initial pheromones are set to record pheromone information. Then, starting from the start node, each node transition is marked to not return the same node to the previous node again, and the next node is switched to the next node based on the pheromone values of the potential nodes that are to be passed next. When it is decided which node is to be preferred, the route with the most pheromones at 50% probability is selected according to the amount of pheromones in the roads. One of the pheromones in the roads that can possibly be travelled with a probability of 50% is decided by the Monte Carlo method. So the ants always search for other possible routes with a probability of 50%, not the way that the pheromone is very much. If a shorter route is found, the pheromone amount is increased excessively according to the definition of global pheromone update in order to make the roads of this route attractive by the end of iteration if the route has a better solution than all other routes. The routes through which all ants complete their tours are increased according to the definition of local pheromone update. Thus the ants will prefer the way in which the tours are completed in the subsequent iterations. The shortest roads in the previous solution will be more attractive than the other roads. In the case of repeating the same lap in each iteration, a large part of the ants will start to prefer this route after a certain time (iteration) and thus the most suitable possible solution will be found.

In situations where there are too many nodes in vehicle locating problems (e.g. 600 nodes), the trial solutions are incredibly high. It is unlikely to find a solution through a non-intuitive method by trying all the ways. In short, it is impossible to solve the optimal solution with the present technological possibilities, or even with less knots; the solution is still practical because it cannot be found in a short time. For these reasons it is much easier to find a solution with an intuitive method. Even if this solution is not the best solution, it is probably the best solution. The ant colony algorithm is one of the most successful algorithms in this respect.

The algorithm is shown below:

1. Start
2. defineAnt_Count, pheromone, pheromone_evaporation, iteration_number etc

3. Define cluster for nodes and node connections
4. while iteration_number > 0
5. while ant_count > = 0
6. randomly generate a real number from 0 to 1
7. If the generated random number is less than q0, select the node with more pheromone density. If not, collect all the pheromone amounts in the alternative pathway and divide the amount of pheromone in each path into the total amount found. Collect alternative paths cumulatively according to the Monte Carlo method and generate a random number from 0 to 1 and choose the node that corresponds to the range.
8. store node and other information in memory
9. If tour is not completed, go to step 6
10. save solution as local solution
11. save as global solution if solution found is better than before
12. loop (step 5)
13. loop (step 4)
14. display the global solution as the best solution
15. End.

Flowchart of ACO application via C# is shown in Figure 2.1 and Figure 2.2.

AGV is usually used to transport goods from one spot to another in factory environments. These tools have predetermined roots.

In this study, rovers of AGV were identified and determined quickly and practically via ant colony optimization. Non-intuitive methods can also be used to determine the routing, but when the number of nodes is large, the number of operations is very large, so heuristic methods are more practical.

The problem is handled as a traveling salesman problem. The AGV which has been released from the depot has been solved by returning to all the stations once and returning to the depot as the last stop.

In the study, stations and distance information of AGV coming out of the depot are shown in the table below. Distances of stations and warehouse are shown in Table 2.1 as meters.

An ant colony optimization program was developed in C # programming language. The machine was an Intel i5 2.4GHz processor with 4GB RAM and a Windows 10 operating system.

In the study $\alpha = 1$ and $\beta = 1$ were taken. The α value indicates the importance of pheromone density. The β value also shows the relative importance of the road. $\rho = 0.1$ was taken. The ρ value is the pheromone evaporation rate. $q_0 = 0.5$. The q_0 value indicates that the ants will choose the path with 0.5 probability of the pheromone being dense. Depending on the amount of pheromone, with a probability of (1 – 0.5), according to the Monte Carlo

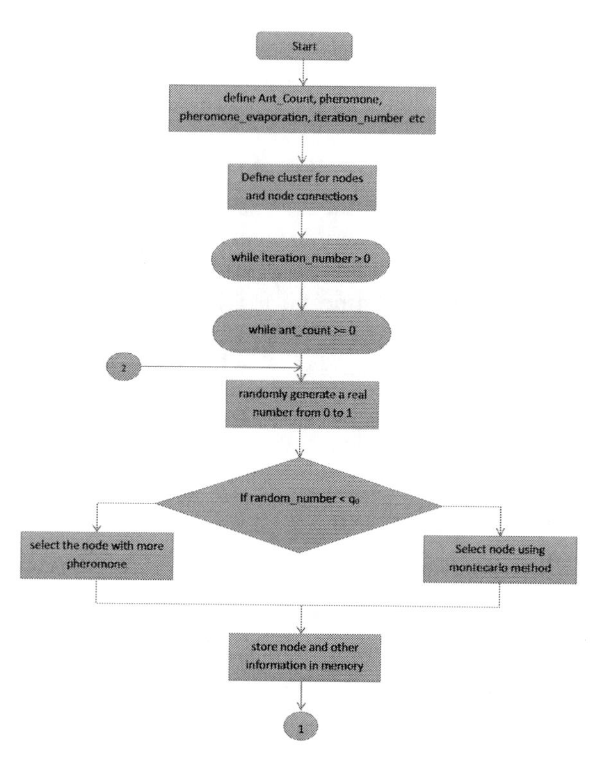

FIGURE 2.1
Flowchart of ACO Application Via C# – 1

method. The number of ants is ten. In the tests performed on the developed program, an average of 1,000 tests takes 2.5 seconds.

According to the results obtained in 1,000 experiments;

Shortest way is 257.61 meters.

Shortest way:

Warehouse → Station 2 → Station 5 → Station 4 → Station 9 → Station 6 → Station 7 → Station 8 → Station 3 → Station 1 → Warehouse

2.5 Conclusion

In this study, the shortest way for the AGV line was found using ant colony optimization. The application is developed in C# programming

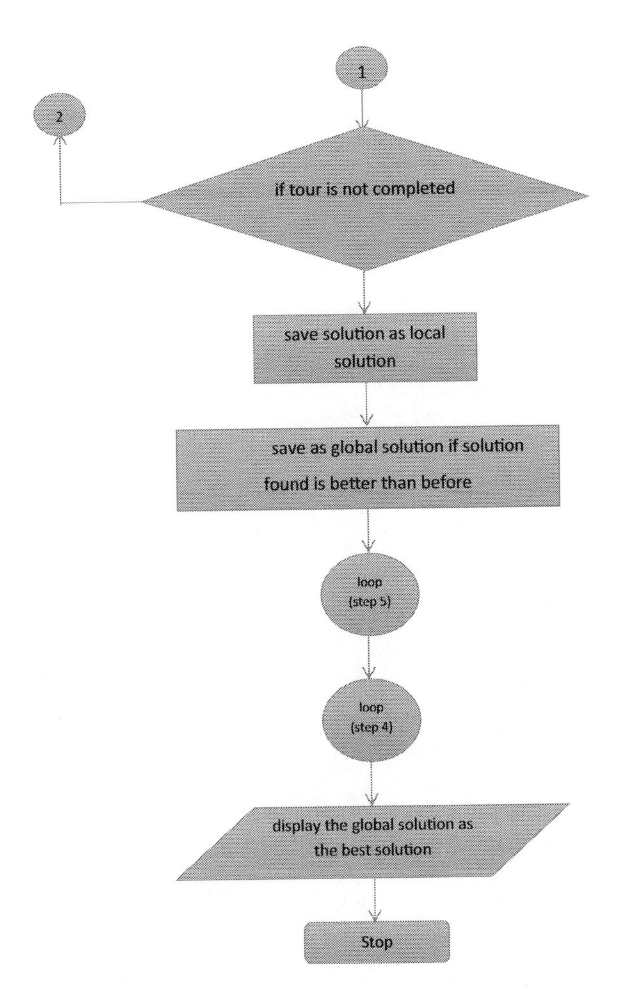

FIGURE 2.2
Flowchart of ACO Application Via C# – 2

language. The ant colony optimization technique is used for multiple AGVs task allocation and path planning problems. As a result, in this study AGV ant colony optimization helped to determine and direct the route more practically and quickly. The advantage of this method arises especially when the number of nodes is large. It can be said that ACO is an effective optimization method for solving any problem that includes the shortest way.

TABLE 2.1

Distances of Stations and Warehouse (Meters)

	WAREHOUSE	STATION 1	STATION 2	STATION 3	STATION 4	STATION 5	STATION 6	STATION 7	STATION 8	STATION 9
WAREHOUSE	0	10	20	20	17.32	31.62	14.14	41.23	51.96	51
STATION 1	10	0	30	17.32	28.28	44.72	22.36	31.62	36.06	50
STATION 2	20	30	0	28.28	17.32	22.36	44.72	60.82	58.31	58.31
STATION 3	20	17.32	28.28	0	41.23	50	44.72	50	31.62	70.71
STATION 4	17.32	28.28	17.32	41.23	0	20	30	50.99	64.03	36.06
STATION 5	31.62	44.72	22.36	50	20	0	50	70.71	78.1	50
STATION 6	14.14	22.36	44.72	44.72	30	50	0	22.36	41.23	31.62
STATION 7	41.23	31.62	60.82	50	50.99	70.71	22.36	0	41.23	50
STATION 8	51.96	36.06	58.31	31.62	64.03	78.1	41.23	41.23	0	82.46
STATION 9	51	50	58.31	70.71	36.06	50	31.62	50	82.46	0

References

Guillaume, D., Michael, D., Abdelghani, B., Damien, T., & Mohamed, D. (2017). Decentralized motion planning and scheduling of AGVs in FMS. *Transactions on Industrial Informatics*. September. http://dx.doi.org/10.1109/TII.2017.2749520

Hamed, F., & Mohammad, S. M. (2015). Methodologies to optimize automated guided vehicle scheduling and routing problems: A review study. *Journal of Intelligent & Robotic Systems, 77*, 525–545. doi:10.1007/s10846-013-0003-8.

Kulatunga, A. K., Liu, D. K., Dissanayake, G., & Siyambalapitiya, S. B. (2006). Ant colony optimization based simultaneous task allocation and path planning of autonomous vehicles. *IEEE Conference on Cybernetics and Intelligent Systems*, Bangkok, Thailand.

Leite, L.F.V., Esposito, R.M.A., Vieira, A. P., & Lima, F. (2015). Simulation of a production line with automated guided vehicle: A case study. *Independent Journal of Management & Production (IJM&P), 6*(2) (April–June).

Marco, D. (2005). BLUM Christian, Ant colony optimization theory: A survey. *Theoretical Computer Science, 344*, 243–278.

Negahban, A., & Smith, J. S. (2014). Simulation for manufacturing system design and operation: Literature review and analysis. *Journal of Manufacturing Systems, 33*, 241–261.

Neradilová, H., & Fedorko, G. (2016). The use of computer simulation methods to reach data for economic analysis of automated logistic systems. *Economic Analysis of Automated Logistic Systems, 6*, 700–710.

Udhayakumar, P., & Kumanan, S. (2010). Task scheduling of AGV in FMS using non-traditional optimization techniques. *International Journal of Simulation, 9*(1), 28–39.

Vavríka, V., Gregora, M., & Grznár, P. (2017). Computer simulation as a tool for the optimization of logistics using automated guided vehicles, *Procedia Engineering, 192*, 923–928.

Wang, Y-C., Chen, T., Chiang, H., & Pan, H-C. (2016). A simulation analysis of part launching and order collection decisions for a flexible manufacturing system. *Simulation Modelling Practice and Theory, 69* (December), 80–91.

3

Data Envelopment Analysis: Applications to the Manufacturing Sector

Preeti

Research Scholar, Department of Management, Birla Institute of Technology, Mesra, Ranchi, India. Preetimathotia@gmail.com

Supriyo Roy

Associate Professor, Department of Management, Birla Institute of Technology, Mesra, Ranchi, India. Supriyo.online@gmail.com

CONTENTS

3.1 Introduction....34
3.2 Data Envelopment Analysis....35
3.2.1 Basic Model and Mathematical Formulation....35
3.2.2 Advanced DEA Models....36
3.2.3 Selection of DEA Model....37
3.3 Literature Review....37
3.3.1 Studies Relating to Performance Measurement of the Manufacturing Sector....38
3.3.2 Studies Relating to Emerging Manufacturing Paradigms....43
3.3.2.1 Advanced Manufacturing Technologies....43
3.3.2.2 Flexible Manufacturing System....45
3.3.2.3 Production Layout and Planning....45
3.3.2.4 E-commerce Innovation in Manufacturing....46
3.3.3 Studies Related to Environmental Impact of the Manufacturing Sector....47
3.3.3.1 Energy Efficiency....47
3.3.3.2 Undesirable Outputs....48
3.3.4 Section Summary....49
3.4 Conclusion and Future Scope....50
References....51

3.1 Introduction

Manufacturing sector is regarded as an integrated model with a number of levels from machines to production to entire business level operations. The sector has evolved considerably compared to the traditional manufacturing which meant conversion of raw materials to finished products to be distributed in the market (Esmaeilian et al. 2016). Developed manufacturing processes and technologies include advanced manufacturing technologies (AMT) like group technologies, flexible manufacturing systems (FMS), industrial robots, computer-integrated manufacturing systems and sustainable manufacturing.

AMT incorporates highly sophisticated and automated computerized design and operations system. The manufacturing sector, using AMT, aims to manufacture better quality products at minimum costs with shortest delivery time using highly sophisticated manufacturing operations. Some examples of AMT technologies are automated material handling systems, industrial robotics, computer numerical control machines, group technology, computer-integrated manufacturing systems (CMS) and flexible manufacturing systems. As per Wang and Chin (2009), FMS enhances manufacturing effectiveness by providing opportunities to manufacturers to perk up their profitability, competitiveness and technology through a highly focused and efficient approach. However, an enterprise resource planning system offers an integrated and comprehensive view of the overall organization's businesses that is helpful for the management of the organization. Application of group technology principles in comparison to the traditional layout in production is one of the recent technologies used in manufacturing organizations. Group technology (GT) refers to arranging machines into different manufacturing cells generating families of groups with similar processing or shapes requirements. One example of the GT principle is a cellular manufacturing system where the layout of the product's production is based on processing requirements. An efficient material handling system plays a serious role in improving the productivity and flexibility of the entire manufacturing system. Integration of the material handling system with other components of manufacturing has been an area of interest to researchers over past decades. CMS is one of the group technologies used for efficient manufacturing. An efficient CMS permits production of diverse products in batches of small lot sizes of minimum annual volumes of production.

Apart from adopted advanced technologies, the manufacturing sector consumes a considerable amount of energy in its various processes. There is an urgent need to optimize the energy use and increase energy efficiency in manufacturing processes and production. Hence, there is increased use of technologies in the manufacturing system to improve energy efficiency. DEA is popularly used by researchers to measure the energy efficiency of the manufacturing system and suggest measures to improve the production processes using advanced technologies. Any manufacturing activity produces both desirable (or good) outputs and undesirable (or bad) outputs. Undesirable outputs can be in form of solid wastes, wastewater or waste gases. A performance evaluation framework of any manufacturing processes must incorporate undesirable outputs for complete result.

The increasing application of data analysis to quantify the performance of manufacturing systems has been widely considered in the previous literature. Traditionally, the analysis of performance measurement was carried out using the univariate ratio approach. This approach suffered from certain limitations. Researchers started exploring the techniques similar to DEA, which is not dependent on the functional relationship of multiple input-output variables. It is one of the techniques recommended for quantifying the operational efficiency of manufacturing systems. The application of DEA in the manufacturing sector expands to include diverse areas, for example choosing the best plant layout, evaluating the most effective FMS and AMT, identifying the most sustainable manufacturing system, finding the energy efficient manufacturing unit, investigating the profitability and marketability of manufacturing organizations.

The chapter is segregated into four sections. DEA methodology in detail is elaborated in section 2. Section 3 covers the body of literature showing application of DEA in manufacturing sector. Section 4 provides the conclusion and future perspective.

3.2 Data Envelopment Analysis

3.2.1 Basic Model and Mathematical Formulation

Charnes et al. (1978) pioneered DEA to calculate the efficiency score of decision-making units (DMUs), on the foundations of frontier efficiency concept proposed by Farrell (1957). DEA, a non-parametric technique, is used for calculating the relative efficiency of comparable DMU. Unlike other approaches, DEA can incorporate multiple inputs-outputs to determine the performance of DMUs. Compared with parametric technique, DEA does not require the input-output functional relationship form. It identifies the efficient DMUs using the mathematical linear programming technique by constructing the frontier for identifying the efficient units and provides improvement for the inefficient units. Hence it is a powerful tool for optimization and measurement of performance of comparable DMU. Generally the efficiency is determined by DEA as:

$$\text{Efficiency} = \frac{\textit{weighted sum of outputs}}{\textit{weighted sum of inputs}}$$

The above computed ratio gives the single relative efficiency measure. DMU with a ratio equal to 1 represents the efficient DMU, whereas a ratio less than 1 indicates less efficient DMU. DEA determines the set of weights that optimizes each DMU and finds the best performing DMU.

CCR (Charnes, Cooper and Rhodes) and BCC (Banker, Charnes and Cooper) models are the two conventional DEA models. The CCR model formulated by Charnes et al. (1978) is also known as the constant return to scale model, whereas the BCC model extended by Banker et al. (1984) is also referred to as the variable return to scale model. In addition, the basic models are either output-oriented or input-oriented. The model that tries to minimize inputs with the given level of output is referred to as an input-oriented model. However, an output-oriented model tries to maximize output with the given level of input.

The application of DEA is wide ranging, such as banks, hospitals, manufacturing firms, telecommunications, schools and non-profit organizations. However, the top five application fields include agriculture, banking, supply chain, transportation and public policy (Emrouznejad and Yang 2018). Apart from basic models of DEA, the other models of DEA include the slack-based model (SBM), additive model, multiplicative model, assurance region DEA model and cone-ratio DEA model. SBM directly finds any shortfalls and excesses in outputs and inputs respectively. An SBM-efficient unit needs to have efficiency score of 1 with no input excesses and output shortfalls. The assurance region model allows flexibility of unrestricted weights for inputs and outputs. Restricted weights help in effective identification of efficient and inefficient units.

3.2.2 Advanced DEA Models

Certain advanced models are developed as an expansion to basic models of DEA.

Super efficiency model: Due to deterministic nature of DEA, any observation with extreme value clearly manipulates the efficiency score. In an effort to minimize the mentioned problem, the super efficiency approach developed by Banker and Chang (2006) is used. It identifies and then removes extreme observations from the sample to achieve a consistent technology frontier.

Network and dynamic model: The conventional or basic models of DEA are regarded as a black box that receives outputs and inputs variables to generate the relative efficiency score. The recent development of dynamic DEA and network DEA models is like opening the black box of conventional models and taking into consideration the intermediating variables.

Fuzzy DEA: Traditional DEA models use certain data for outputs and inputs. However, in presence of uncertain data involving vagueness, the fuzzy DEA model introduced by Sengupta (1992) is used. Implementation of fuzzy logic in traditional DEA models solves the problem of ambiguous data in performance measurement problems.

Hybrid model: The hybrid model is an integrated approach of using more than one techniques for performance measurement of any production unit. The analytic hierarchy process (AHP) is among the popularly accepted extensions to the DEA model. AHP analyzes qualitative criteria for making the most strategic decisions based on expert judgments (Saaty 2008). A hybrid model of AHP and DEA can be utilized for multiple criteria techniques for evaluating performance. Similarly, a hybrid model of DEA with parametric technique like stochastic frontier analysis (SFA) is also found in previous literatures. SFA is an econometric model given by Aigner (1977) and Meeusen and van Den Broeck (1977). Simulation is yet another technique employed by researchers with DEA for performance measurement. Since manufacturing sectors depicts a complex system, the application of simulation will provide a system-wide view of new technology before the actual execution in the processes (Shang and Sueyoshi 1995).

Bootstrapping approach: Simar and Wilson (2007) proved that DEA-based studies using a two-stage performance evaluation process has the problem of correlation between the estimated efficiency score. Hence, Simar and Wilson developed a bootstrap procedure based on a re-sampling technique that removes the problem of correlation in efficiency estimates.

3.2.3 Selection of DEA Model

The selection of the DEA model involves not only model selection. The analyst has to make decision regarding return to scale, model orientation and type of outputs (desirable or undesirable). DEA model can be implemented using single stage or multiple stage using intermediate variables. Figure 3.1 shows the logical steps involved in DEA model implementation.

The DEA model starts with identification of the problem. The next step involves arranging the data to be considered as input and output variables. The best fitted DEA model is applied for data analysis and the best optimal decision is made.

3.3 Literature Review

DEA is applied to different aspects of manufacturing sector. Hence, the literature review of manufacturing sector is classified into four different areas. They are (1) studies related to performance measurement of manufacturing sector, (2) studies related to different manufacturing paradigms and (3) studies related to environmental impacts of manufacturing.

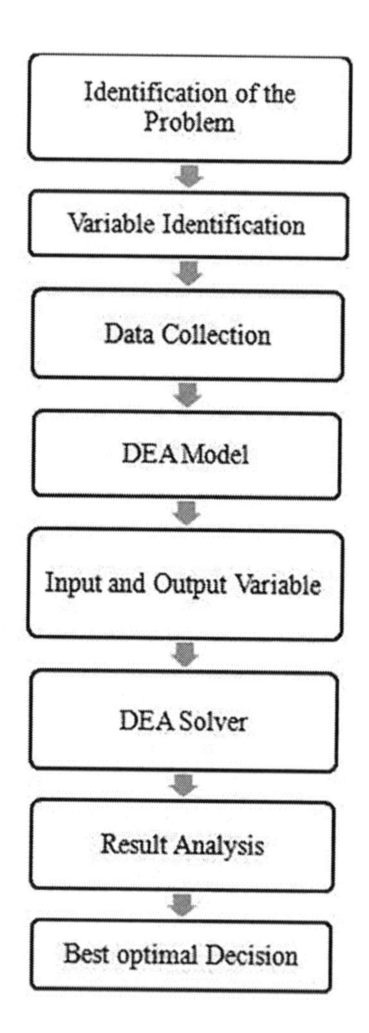

FIGURE 3.1
Steps for DEA model

3.3.1 Studies Relating to Performance Measurement of the Manufacturing Sector

DEA, a mathematical linear programming approach, is utilized for performance measurement, estimation, optimization and benchmarking. It is a simple technique employed for identifying the best and worst performers. Mahadevan (2002) estimated the performance in terms of productivity growth of manufacturing sector in Malaysia. The study calculated productivity growth using the Malmquist index (MI) to identify crucial sources of productivity growth for policy formulation. Talluri et al. (2003) analyzed the transformation from manufacturing to business performance in the auto supplier industry using the non-parametric decision-making and optimization technique referred to as DEA. The study tried to examine the performance of supplier. Sheu and Peng (2003) assessed a manufacturing plant in Taiwan to develop a complete manufacturing management model to benchmark the performance indices in factory, line and station level. The study used DEA for evaluating efficiency and productivity by employing production as outputs, labors and the number of key machines employed as inputs. Braglia et al. (2003) used DEA to estimate the performance of five steel plants of a large private company in Italy from 1995 to 1997. Reduction of factors, cross-efficiency and stepwise approaches is used to advance the discriminating power of DEA. The critical investigation of the performance of each plant helps the management of the company to formulate manufacturing strategies.

Narasimhan et al. (2004) developed a two-stage conceptual model using two new constructs to link manufacturing investment and manufacturing performance. Flexibility and execution competence are the two new intervening variables that are relevant to manufacturing firms. Hence the developed model successfully links flexibility to performance in the manufacturing sector. The analysis of result highlights firms that effectively convert flexibility capabilities into superior levels of performance. Cook and Green (2004) used DEA to recognize the efficient business activity for a company that operates within several plants in the rolled steel sector. The modified version of DEA is used instead of conventional DEA model. Decisions regarding improvement, de-emphasis or phasing out of any business component can be taken by management of the steel company. Leachman et al. (2005) constructed composite performance metrics for measuring firms manufacturing competitiveness. The study identified three practices as important for superior manufacturing. These key practices include time compression at time of production, R&D commitment and degree of outsourcing. Düzakın and Düzakın (2007) performed efficiency analysis of companies from 500 major companies in Turkey. Super-SBM of DEA is employed using two inputs and three outputs. The study suggested not reducing the net assets or number of employees to increase efficiency, rather recommending to solve the problem of inefficiency by concentrating on increasing outputs.

Mok et al. (2007) assessed the significance of leverage in enhancing the firm's performance by measuring technical efficiency using DEA for foreign-invested toy manufacturing firms. The second stage analysis of efficiency score reveals the positive relation between profitability and efficiency of the firm. Azadeh et al. (2007) developed an integrated framework using a PCA and DEA approach to measure and rank manufacturing units on the basis of equipment performance indicators. The indicators identified are linked to equipment efficiency, productivity, profitability and effectiveness. Further, the results arrived are validated and verified using an NT approach and

Spearman and Kendall-Tau tests. The result of the study identified indicators that have considerable effects on the manufacturing sector's performance.

Saranga (2009) investigated the auto component industry to test the cost efficiency of domestic suppliers by using advanced models of DEA. The study showed that substitution of labor for capital is the main source of inefficient performance in the Indian component industry. The study suggested the need to identify those labor law reforms which contribute to inefficiency.

In the time of globalization, the United States manufacturers are outsourcing ample portions of their businesses to small and medium-sized manufacturing enterprises (SMEs). Thus, the SMEs have to compete with global competitive market to survive the competition. Ahmad and Qiu (2009) designed an integrated AHP/DEA productivity-oriented model for 3,000 SMEs by linking both qualitative and quantitative factors to assess the performance. The integrated model is better in analyzing the broad variety of problems compared to the single-model-based tool. Zeydan and Çolpan (2009) combined DEA and fuzzy technique for order preference by similarity to ideal solution (TOPSIS) to incorporate both qualitative and quantitative data for performance measurement and evaluation. The study aimed to identify the most efficient job shops in Turkey.

Many studies have ignored the dynamic effects while measuring the performance within the production network. A manufacturing system comprises many production networks having dynamic effects. Using conventional DEA could provide a biased result by ignoring this dynamic effect. Chen (2009) proposed to fit the dynamic effect into the efficiency measurement by developing a network DEA model. The proposed approach is built upon an interrelated and hierarchical production structure within manufacturing units. An overall production system comprises three subcomponents, namely demand support, production design and operations. Lee and Johnson (2011) employed network DEA and MPI to determine the efficiency of the overall production system of the semiconductor manufacturing industry. The result highlighted that changes in productivity is mainly due to demand fluctuations rather than by technical changes in production capabilities.

Halim (2010) undertook to benchmark marketing productivity for the time period of 2001 to 2007 and suggested areas for improvement. The study evaluated the performance of manufacturing firms listed on the Indonesian Stock Exchange. Liu et al. (2010) estimated the operational efficiency of thermal power plants in Taiwan. The study suggested that decline in electricity consumption proves to be the helpful means to improve the operation of the ineffective utilities. Effective performance and realistic target setting are of paramount importance for manufacturing companies. Jain et al. (2011) assessed the performance of two manufacturing organizations involved in distinct part production. Multiple models of DEA are developed for two manufacturing organizations and the most suitable model with desired discriminatory power is selected. The analysis helps to identify the performance improvement targets and its comparisons with peers. Lee and Pai (2011) employed DEA to measure the operational efficiency of top 10 TFT-LCD large-scale manufacturers from 2002 to 2007.

In recent times, a firm's productivity and efficiency depend to a great extent on an efficient intellectual capital management. Costa (2012) used DEA and MPI to measure the best practices implemented for managing intellectual capital and its relationship with performance of Italian yachting companies. Halkos and Tzeremes (2012) evaluated the efficiency of Greek manufacturing sectors with bootstrap techniques using financial data. The first-stage result proved that the efficiencies calculated are biased. After application of the bootstrap technique, the study claimed that efficiency scores have been improved significantly. Shrivastava et al. (2012) investigated the technical performance of coal power plants using DEA models. The findings of the study highlighted reasons for poor performance of inefficient plants.

Ayhan et al. (2013) presented a novel approach to monitor and measure innovation within manufacturing system on the basis of four most critical components used as basic activities. The four important components to measure the degree of innovation are average labor utilization, unit production time, cumulative bottleneck ratio and unit production cost. The developed approach is verified with its applications to manufacturing industry. Mitra Debnath and Sebastian (2014) developed output-oriented DEA framework for Indian steel manufacturing industries to measure its performance. The study points out that location disadvantage condition is due to scattered location of public sector undertakings (PSUs) in India. The analysis of the study showed 45% of private manufacturing units are found to be technical as well as scale inefficient. Park et al. (2014) undertook the performance measurement of the ship block manufacturing industry. The study used process mining and DEA to develop a block manufacturing process (BMP) performance evaluation approach.

There is a need to measure the productivity of production lines on a timely basis to improve productivity. Using one-dimensional indicators does not provide sufficient information about the improvement required. Hence, the application of a number of initiatives related to productivity improvement is required to attain the desired results. Jagoda and Thangarajah (2014) employed DEA approach to analyze productivity improvement initiatives employed by the Canadian packaging industry. The analysis revealed the simultaneous usage of all options provides the highest productivity.

Every organization undertakes continuous evaluation of their operations strategy in comparison to the rivals. Wang et al. (2015) measured the operational performance of 281 listed electronic firms in Taiwan using the dynamic DEA model. The basic objective was to analyze whether the performance of firms improves after recognizing the asset impairments within the firm. The analysis of the result revealed that timely recognition of asset impairments by the managers will result in increased performance by the firms. Abbasi and Kaviani (2016) built a framework for analyzing the performance of cement manufacturers of Iran on the basis of their operations strategies. The study used DEA models comprising fuzzy DEA (FDEA), Grey DEA and imprecise DEA (IDEA) to calculate the operational efficiency of the organizations.

Wang and Chien (2016) proposed a novel framework to diagnose the performance of Taiwanese LED manufacturers. The effect of operational efficiency on the performance of manufacturers is studied using a balanced scorecard in combination with DEA. The financial and non-financial indicators are also included in the performance measurement framework. Tsaur et al. (2017) proposed a four-stage approach, including DEA, MPI, Grey relation analysis and the entropy method, to verify the performance of TFT-LCD companies in Taiwan. Lozano et al. (2017) developed DEA model to find out the cost and allocative and technical efficiency of a reconfigurable manufacturing system. Table 3.1 provides a concise summary of the above-mentioned studies.

TABLE 3.1

Summary of DEA application for measuring performance of manufacturing sector

Author and year	Application field
Mahadevan (2002)	Productivity growth of 28 industries in Malaysia's manufacturing sector
Talluri et al. (2003)	Transformation from manufacturing to business performance in the auto supplier industry
Sheu and Peng (2003)	Manufacturing management model for notebook manufacturing plant in Taiwan
Braglia et al. (2003)	Performance of steel plants of large private company in Italy
Narasimhan et al. (2004)	Manufacturing sector performance by linking flexibilities to performance
Cook and Green (2004)	Efficient business activity within several plants in rolled steel sector
Leachman et al. (2005)	Performance of automobile industry in contrast to its competitors
Düzakın and Düzakın (2007)	Efficiency analysis of companies in Turkey
Mok et al. (2007)	Efficiency analysis of foreign toy manufacturing firms
Azadeh et al. (2007)	Integrated framework of DEA and PCA to rank manufacturing units on the basis of equipment performance indicators
Saranga (2009)	Performance measurement of auto component industry in India
Ahmad and Qiu (2009)	Performance of SME and establish an efficient benchmark in the SME manufacturing enterprises
Zeydan and Çolpan (2009)	Efficiency analysis of job shops in Turkey
Chen (2009)	Performance improvement dynamic production network within manufacturing system
Halim (2010)	Marketing productivity of manufacturing firms in Indonesia

(Continued)

Author and year	Application field
Liu, Lin and Lewis (2010)	Operational performance of thermal power plants in Taiwan
Jain et al. (2011)	Performance and benchmarking of manufacturing systems
Lee and Johnson (2011)	Efficiency of overall production system of semiconductor manufacturing industry
Lee and Pai (2011)	Operational performance of TFT-LCD large scale manufacturer
Costa (2012)	Productivity and efficiency of intellectual capital
Halkos and Tzeremes (2012)	Performance of Greek manufacturing sectors
Shrivastava et al. (2012)	Technical efficiency of coal-fired power plants
Ayhan et al. (2013)	Innovation efficiency in manufacturing processes
Mitra Debnath and Sebastian (2014)	Scale efficiency in the manufacturing firms
Park et al. (2014)	Performance of the ship block manufacturing processes in the ship building industry
Jagoda and Thangarajah (2014)	Productivity improvement initiatives used by Canadian packaging industry
Wang et al. (2015)	Operational performance of electronic firms in Taiwan
Abbasi and Kaviani (2016)	Operational performance evaluation framework cement industry in Iran
Wang and Chien (2016)	Operational performance of Taiwanese LED manufacturers
Tsaur et al. (2017)	Operational performance of TFT-LCD companies in Taiwan
Lozano et al. (2017)	Cost and allocative and technical efficiency of a reconfigurable manufacturing system

3.3.2 Studies Relating to Emerging Manufacturing Paradigms

The application of DEA in manufacturing sector has been studied in diverse areas, for example, selecting the best plant layout and evaluating the most effective FMS and AMT and ecommerce innovation in manufacturing processes. As different models are being applied to varied manufacturing paradigms, the existing models and upcoming models of DEA need to be studied carefully to explore the performance of new manufacturing technologies.

3.3.2.1 Advanced Manufacturing Technologies

Khouja (1995) build a two-stage decision model to address the problems related to selection of manufacturing technology. DEA is used in the first phase to identify the efficient technology with optimal combination of performance parameters. In the next stage, multi attribute decision-making is used to select among the technologies identified as efficient in phase 1. The real dataset of 27 industrial robots is used to illustrate the proposed model. Baker and Talluri (1997) aimed to improve the work of Khouja by introducing the concept of cross-efficiency for selection of robots. Karsak (1998) developed a two-stage approach for selection of robots. In the first phase, DEA is employed to identify the efficient robot alternatives. In the second stage, a fuzzy algorithm is developed to position the best robot alternatives. The fuzzy algorithm considers vendor-related subjective criteria for ranking the alternatives. Hence, the decision-makers are able to finalize the ranking of robots considering both the engineering costs as well as the subjective criteria. The algorithm developed in the study can be applied to diverse areas like determining the best CNC machine or FMS and selecting the efficient facility selection. Braglia and Petroni (1999) also used DEA for selecting the best industrial robots. The study aimed to identify the most optimal robot by measuring the relative efficiencies of 12 robot manufacturers. The study applied cross-efficiency approach of DEA for comparison of results.

AMT helps in taking complex decisions that requires cautious consideration of diverse performance criteria. Talluri and Yoon (2000) utilize a cone-ratio DEA model for the AMT selection process. The study uses a real dataset of industrial robots to depict the proposed model. The initial step of the study identifies a feasible set of AMT which meets the system requirements and budget constraints. The competitive variables like quality, cost, time and flexibility are defined and compared to the selected performance variables to identify the preference relationships. The alternative system selected is evaluated based on preference relationships and performance criteria. Amin and Emrouznejad (2007) improved the multi-criteria decision-making (MCDM) DEA model developed by Karsak and Ahiska (2005) for identification of the best AMT. The paper eliminated the need to solve the linear programming by offering the compact form to find out the maximum value of ε, non-Archimedean that improves the discrimination among the DMUs. This improved approach is beneficial when computational complexity analysis is to be carried out. Wang and Chin (2009) applied double frontier DEA model for identifying the best and worst relative efficiencies of AMTs. Advanced model of DEA that is used is proved to be better in identifying the best AMTs compared to the traditional models used. The paper showcases four different examples to better illustrate the ease of using DEA in AMT evaluation.

Worker assignment is an important part of cell manufacturing, which is an old hurdle in the cellular manufacturing system. Ertay and Ruan (2005) concentrated on the labor assignment problems by selecting the most suitable operator allocation in the cellular manufacturing system environment. DEA based framework is developed to identify efficient and inefficient labor assignment alternatives from the dataset comprising 48 simulation scenarios. The DEA approach employed the average operator utilization, the average lead time as the output variables and transfer batch size, demand level and number of operators as the input variables. Ten labor assignment alternatives out of 48 simulation scenarios are found to be relatively efficient. Azadeh et al. (2015) proposed to develop a model considering the decision-making style of operators that acts as a representative of an operator's personnel attributes that increases the productivity and satisfaction of the system. The point is to minimize the costs of intracellular movement and cell establishment and also to minimize the inconsistency in decision-making styles among operators. The paper used the $\mathcal{E}$-constraint method and multicriteria decision analysis (MCDA) DEA to solve and gather Pareto optimal solutions and to choose the top solution among the selected solutions. The study justified that proposed model will be useful to design CMS with more operators' productivity and satisfaction.

3.3.2.2 Flexible Manufacturing System

Shang and Sueyoshi (1995) developed an integrated framework using AHP, simulation, accounting and DEA for FMS design, planning and selection. AHP quantifies qualitative and long-term benefits at the strategic level. Medium-term tactical impacts are determined using simulation. Accounting modules identifies inputs like costs and resources. DEA combines these three modules (qualitative, quantitative and financial) to identify the best FMS. Sarkis (1997) used basic and extension models of DEA for the decision-making process regarding investment in FMS. Variables considered for the study includes total cost, throughput time, labor requirements, work in progress, volume flexibility, space requirements, process/routing flexibility and product-mix flexibility. Hence the study aimed to include both traditional and non-traditional factors in the decision-making process. Petroni and Bevilacqua (2002) employed a DEA-based approach to identify the best firm with highest manufacturing flexibility capabilities. Data is collected via questionnaire containing seven basic aspects of manufacturing flexibility. On-site interviews and discriminant analysis are also carried out to analyze the organizational and contextual variables that characterize to reach the best practice firm status. Liu (2008) developed a fuzzy DEA/AR model for the performance measurement of flexible manufacturing system. Since the input and output data are characterized as fuzzy data, the fuzzy DEA model is used. Input data includes operating costs, capital, and floor space requirement. Output data includes work-in-process (WIP) improvements, qualitative factor, yield and numbers of tardy jobs. The study found the most efficient FMS amongst the 12 FMS alternatives. Jahanshahloo et al. (2009) attempted to correct the model and proof of theorem developed by Liu (2008) for FMS selection. They study presented corrected model for evaluating the lower and upper bound for solving fuzzy DEA model with restricted multiplier. Zhou et al. (2010) further corrected the works of Jahanshahloo et al. (2009) and Liu (2008) by proving that both the studies omitted the verification step for constraints of assurance regions.

3.3.2.3 Production Layout and Planning

Successful manufacturing system demands a well-designed plant layout for efficient production and service systems. Metters and Vargas (1999) used

DEA to evaluate the simulation results of master production scheduling dual buffer stock system. Ertay et al. (2006) attempted to design an efficient layout framework by evaluating the facility layout design. DEA is used in conjunction with AHP to solve the layout problems of manufacturing systems. The layout problems of manufacturing processes usually consist of stochastic data. Hence there is a need to extend new approaches that tackles the problems of stochastic data. Azadeh et al. (2015) integrated computer-simulation and stochastic data envelopment analysis approaches to solve layout design problems in the manufacturing processes. The study considers the impacts of environmental and safety issues and the issue of stochastic outputs of job shop facility layout design problems. Azadeh et al. (2010) built an effective and superior model for optimization of operator allocation in CMS. A combined fuzzy DEA (FDEA) computer simulation and fuzzy C-means is used to carry out the optimal selection of operator layout in CMS environment. Al-Refaie (2011) proposed an approach to optimize multiple correlated quality characteristics (QCH) in a robust design that can be adopted for a wide range of manufacturing applications. The proposed approach is developed using principle component analysis (PCA) and DEA. Amirteimoori and Kordrostami (2012) addressed the production planning problems by determining the most favorable production plans using the DEA approach. The model takes the production levels and size of the operational units into consideration to plan to production level. The approach suggests to not setting huge production plans with small inputs and outputs.

3.3.2.4 E-commerce Innovation in Manufacturing

Lately, the manufacturing sector has witnessed the increasing use of information technology like internet-based customer transaction processing to enterprise resource planning system. Ng and Chang (2003) combined DEA and the Cobb-Douglas production function to study the significance of computerization in improving the performance of the Chinese enterprises. The result confirmed the importance of computerization in achieving highest competitiveness in the Chinese enterprises. The analysis of the study shows low level of efficiency of the Chinese enterprises due to failure in utilizing the resources to full potential. The Cobb-Douglas production function model shows improvement in enterprise output can be achieved by using computerization. Investments should be aimed to elevate the level of computerization by increasing investment in the strategic information technology infrastructure. Ross and Ernstberger (2006) investigated the relevance of information technology on productivity of 51 manufacturing firms selected out of 500 IT-enabled firms in US. The study used total workforce size, IT-specific workforce size and IT budget projections as input variables. Market performance (revenue-related) and financial performance (profit-related) have been utilized as output variables indicating firm's performance. The study identified the most efficient firm showing the benchmark performance level.

Wu et al. (2016) employed enhanced DEA by integrating the theory of satisfaction degree for maximizing the optimal weights of DMUs by using cross-efficiency concept. The developed approach is applied in the company for server selection to include enterprise resource planning system. The study considers one input and five outputs to shortlist the best performing server for the company.

3.3.3 Studies Related to Environmental Impact of the Manufacturing Sector

The manufacturing sector consumes a considerable amount of energy in its various processes. At the same time, important by-products of any manufacturing activity include undesirable outputs like waste gases, solid wastes and wastewater. There is an urgent need to optimize the energy use and also consider undesirable outputs to effectively measure the manufacturing processes and production. DEA is popularly used by researchers in measuring the energy efficiency of manufacturing system and suggest measures to improve the production processes using advanced technologies.

3.3.3.1 Energy Efficiency

Due to ever-increasing growth, the energy demand is also increasing across every sector in the economy (transportation, residential, industrial and commercial). There is a need to sustain the energy consumed by all sector of manufacturing industries. Önüt and Soner (2007) found the energy efficiency of medium-sized companies with a number of workers ranging from 100 to 200. The study aimed to measure the energy efficiencies and identify the best and worst performing companies. The study suggested measures to cut down resource consumptions for inefficient companies. Azadeh et al. (2007) employed DEA, numerical taxonomy and PCA, an integrated approach in energy-intensive manufacturing sectors for measurement and optimization. The paper assessed energy consumption to measure improvements required for attaining the benchmark energy efficiency level. Mukherjee (2008) made an effort to investigate the inefficient use of energy in the Indian manufacturing industries and observed the interstate pattern of energy efficiency. DEA model developed used gross value of production from manufacturing industries in the state as output, and energy, capital, labor and materials as input variables. The model result revealed the average amount of energy required by manufacturing units to produce certain amount of energy. The second stage analysis, using regression, explored the relation between output share and energy efficiency in industries.

In addition, several studies considered undesirable outputs into the energy efficiency analysis. Mandal (2010) assessed the energy efficiency considering undesirable outputs of the Indian cement industry. The study proved that without the inclusion of undesirable outputs, the energy estimates are biased. The presence of environmental regulations not only reduces pollution levels but also enhances the energy use efficiency estimates in the Indian cement industry. Wu et al. (2012) assessed energy efficiency of industrial sectors in China using a DEA model with CO_2 emissions. The study measures both dynamic and static efficiency performance indexes with a framework of undesirable and desirable outputs. The study found 82.3% improvement in dynamic energy performance. He et al. (2013) quantified productivity and energy effectiveness of 50 enterprises in iron and steel industry of China using conventional DEA and MI. The study also considered undesirable outputs to measure the productivity change using the Malmquist-Luenberger index. The result proved that environmental regulations such as raising the emission standard and charging a pollution emission tax that increases the environmental costs of the firm has a positive effect on the productivity of China's iron and steel industry. Bi et al. (2014) measured the environmental performance of thermal power generation sector under the environmental regulation framework using the DEA approach. The study used the SBM model of DEA to provide additional insights for environmental protection and energy utilization. The result found low environmental and energy efficiency of the power generation sector. Perroni et al. (2017) applied three different techniques to measure the relative efficiency: DEA, SFA and COLS. The regression quantile is applied to test the link between the enterprise efficiency and the energy efficiency practices adoption level.

3.3.3.2 Undesirable Outputs

Several studies proved that true productivity growth is underestimated if undesirable outputs are ignored. Zaim (2004) included pollution as part of undesirable output of manufacturing activity for developing pollution intensity index using DEA framework. Pollution intensity index measures environmental efficiency in US state manufacturing sectors between 1974 and 1986. Inputs used in the model include capital stock and manufacturing employment aggregates. The good and bad outputs included in the study are gross state product and emissions of sulphur oxides, nitrogen oxides and carbon monoxide by the manufacturing sector. The analysis revealed that contribution of polluting industries in total manufacturing activity and contribution of manufacturing in total state product are the significant factors for pollution intensity changes in the manufacturing industries. Sözen et al. (2010) developed an operational and environmental model to find out the performance of the power plants in Turkey. In the first model, main production indicators and fuel cost per actual production indicators were taken as input and output variables, respectively. In the second model, environmental wastes from power plants are included in the output variables. The analysis identifies the efficient and inefficient power plants. The study proved to be beneficial for Turkey to effectively evaluate the environmental effects and the cost of electricity generation.

Several studies undertook inter-country comparisons of energy efficiency in manufacturing industries. Oggioni et al. (2011) compared the eco-efficiency of 21 countries representing 90% of cement production around the world. The study used DEA and a directional distance approach to construct a framework including both undesirable and desirable outputs to estimate the significance of environmental regulations on the cement industry. Riccardi et al. (2012) studied the influence of carbon dioxide emissions on the performance of cement industries. The comparison is made between traditional industrialized countries and emerging producers like China, India, Brazil and Turkey. DEA models and DDF were used to analyze the efficiency with and without undesirable outputs. The study confirmed that undesirable outputs like carbon dioxide emissions cannot be excluded from the evaluation.

China's economic growth is controlled by manufacturing sector. Hence the Chinese government has identified the significance of sustainable development via energy conservation. Congestion is referred to as an economic process where any reduction in inputs will result in improvement in outputs without any effect on other inputs and outputs. Hence, to estimate the exact performance and proper allocation of resources, it is of prime importance to measure congestion in the industry. Wu et al. (2013) developed a framework to quantify congestion with undesirable outputs in the industry of China. Hence the study recommended that investing fewer inputs in a congested industry will produce more desirable outputs and lesser undesirable outputs. Li and Lin (2016) puts forward an improvement estimation method to determine a green productivity index for China's manufacturing sector. Moreover, the relationship between government measures and green productivity growth is studied. The Malmquist-Luenberger productivity index calculated productivity index accounting for carbon dioxide emissions. The paper proved that China's manufacturing sector requires energy conservation policies to achieve the optimal level of green productivity.

Wu et al. (2017) developed an evaluation approach to measure sustainable manufacturing when wastes from the production are recycled and reused. The study proposed a two-stage network approach instead of the traditional two-stage DEA model considering desirable and undesirable outputs. The Nash bargaining game is used to propose a fair decomposition of efficiency between two stage model of sustainable manufacturing process. The developed model is applied to an iron and steel manufacturing firm in China.

The aforementioned literature shows the trend of studies assessing the environmental impacts of the manufacturing sector. It shows a clear indication of need of more research in this area.

3.3.4 Section Summary

Since the application of DEA in manufacturing sector is diverse, the papers are located in a variety of journals. Table 3.2 provides list of total reviewed articles and their classification in a variety of journals.

TABLE 3.2

Distribution of research paper as per journal

International Journal of Production Research	14
Energy Policy	12
European Journal of Operational Research	7
International Journal of Production Economics	5
Computers and Industrial Engineering	5
Expert Systems with Applications	4
Production Planning and Control	4
International Journal of Operations and Production Management	3
Journal of Operations Management	1
Production and Operations Management	1
Journal of the Operational Research Society	1
Others	15
Total	72

As per the review, *International Journal of Production Research, Energy Policy* and *European Journal of Operational Research* top the list with the highest number of DEA-related publications in the manufacturing sector. *Energy Policy* is in second place because of the considerable number of papers related to energy efficiency and undesirable outputs. The "Others" category in the table sums up all the miscellaneous journals.

3.4 Conclusion and Future Scope

The chapter extends to present the review of published works showing diverse application of DEA in manufacturing sector. The review section is strategically divided to study the DEA application in detail. The DEA technique is explicitly explained, ranging from conventional models to advanced models. Present times are witnessing development in the manufacturing sector in terms of new technologies that enable agility, flexibility and reconfigurability. Bio-manufacturing, advanced materials, semiconductors, nanomanufacturing, additive manufacturing and sustainable manufacturing are some of the emerging manufacturing concepts. DEA can be extensively used to evaluate the new technologies to identify the most efficient alternative. Assessing the environmental impact of these new technologies is also an area where DEA can be utilized successfully. Since the manufacturing sector consumes an ample amount of energy, the researchers in the manufacturing sector can use DEA to establish energy efficiency benchmarks for sustainable achievements. Advanced models of DEA can be used as a modification to the basic model to improve the existing results in manufacturing sector. However, automation in selecting input-output variables for DEA models will considerably enhance the prospect of using DEA in manufacturing. In light of new developments in manufacturing, the scope of applying DEA in solving manufacturing-related problems is wide.

References

Abbasi, M., & Kaviani, M.A. (2016). Operational efficiency-based ranking framework using uncertain DEA methods: An application to the cement industry in Iran. *Management Decision*, *54*(4), 902–928.

Ahmad, N., & Qiu, R.G. (2009). Integrated model of operations effectiveness of small to medium-sized manufacturing enterprises. *Journal of Intelligent Manufacturing*, *20*(1), 79.

Aigner, D. (1977). Formulation and estimation of SFA production function models. *Journal of Econometrics*, *6*(1), 21–37.

Al-Refaie, A. (2011). Optimising correlated QCHs in robust design using principal components analysis and DEA techniques. *Production Planning & Control*, *22*(7), 676–689.

Amin, G.R., & Emrouznejad, A. (2007). A note on DEA models in technology selection: An improvement of Karsak and Ahiska's approach. *International Journal of Production Research*, *45*(10), 2313–2316.

Amirteimoori, A., & Kordrostami, S. (2012). Production planning in data envelopment analysis. *International Journal of Production Economics*, *140*(1), 212–218.

Ayhan, M.B., Öztemel, E., Aydin, M.E., & Yue, Y. (2013). A quantitative approach for measuring process innovation: A case study in a manufacturing company. *International Journal of Production Research*, *51*(11), 3463–3475.

Azadeh, A., Amalnick, M.S., Ghaderi, S.F., & Asadzadeh, S.M. (2007). An integrated DEA PCA numerical taxonomy approach for energy efficiency assessment and consumption optimization in energy intensive manufacturing sectors. *Energy Policy*, *35*(7), 3792–3806.

Azadeh, A., Anvari, M., Ziaei, B., & Sadeghi, K. (2010). An integrated fuzzy DEA—fuzzy C-means—simulation for optimization of operator allocation in cellular manufacturing systems. *The International Journal of Advanced Manufacturing Technology*, *46*(1–4), 361–375.

Azadeh, A., Ghaderi, S.F., & Ebrahimipour, V. (2007). An integrated PCA DEA framework for assessment and ranking of manufacturing systems based on equipment performance. *Engineering Computations*, *24*(4), 347–372.

Azadeh, A., Nazari, T., & Charkhand, H. (2015). Optimisation of facility layout design problem with safety and environmental factors by stochastic DEA and simulation approach. *International Journal of Production Research*, *53*(11), 3370–3389.

Azadeh, A., Rezaei-Malek, M., Evazabadian, F., & Sheikhalishahi, M. (2015). Improved design of CMS by considering operators decision-making styles. *International Journal of Production Research*, *53*(11), 3276–3287.

Baker, R.C., & Talluri, S. (1997). A closer look at the use of data envelopment analysis for technology selection. *Computers & Industrial Engineering, 32*(1), 101–108.

Banker, R. D., & Chang, H. (2006). The super-efficiency procedure for outlier identification, not for ranking efficient units. *European Journal of Operational Research, 175*(2), 1311–1320.

Banker, R.D., Charnes, A., & Cooper, W.W. (1984). Some models for estimating technical and scale inefficiencies in data envelopment analysis. *Management Science, 30*(9), 1078–1092.

Bi, G.B., Song, W., Zhou, P., & Liang, L. (2014). Does environmental regulation affect energy efficiency in China's thermal power generation? Empirical evidence from a slacks-based DEA model. *Energy Policy, 66*, 537–546.

Braglia, M., & Petroni, A. (1999). Evaluating and selecting investments in industrial robots. *International Journal of Production Research, 37*(18), 4157–4178.

Braglia, M., Zanoni, S., & Zavanella, L. (2003). Measuring and benchmarking productive systems performances using DEA: An industrial case. *Production Planning & Control, 14*(6), 542–554.

Charnes, A., Cooper, W.W., & Rhodes, E. (1978). Measuring the efficiency of decision making units. *European Journal of Operational Research,2*, 429–444.

Chen, C.M. (2009). A network-DEA model with new efficiency measures to incorporate the dynamic effect in production networks. *European Journal of Operational Research, 194*(3), 687–699.

Cook, W.D., & Green, R.H. (2004). Multicomponent efficiency measurement and core business identification in multiplant firms: A DEA model. *European Journal of Operational Research, 157*(3), 540–551.

Costa, R. (2012). Assessing intellectual capital efficiency and productivity: An application to the Italian yacht manufacturing sector. *Expert Systems with Applications, 39*(8), 7255–7261.

Düzakın, E., & Düzakın, H. (2007). Measuring the performance of manufacturing firms with super slacks based model of data envelopment analysis: An application of 500 major industrial enterprises in Turkey. *European Journal of Operational Research, 182*(3), 1412–1432.

Edy Halim, R. (2010). Marketing productivity and profitability of Indonesian public listed manufacturing firms: An application of data envelopment analysis (DEA). *Benchmarking: An International Journal, 17*(6), 842–857.

Emrouznejad, A., & Yang, G. L. (2018). A survey and analysis of the first 40 years of scholarly literature in DEA: 1978–2016. *Socio-Economic Planning Sciences, 61*, 4–8.

Ertay, T., & Ruan, D. (2005). Data envelopment analysis based decision model for optimal operator allocation in CMS. *European Journal of Operational Research, 164*(3), 800–810.

Ertay, T., Ruan, D., & Tuzkaya, U.R. (2006). Integrating data envelopment analysis and analytic hierarchy for the facility layout design in manufacturing systems. *Information Sciences, 176*(3), 237–262.

Esmaeilian, B., Behdad, S., & Wang, B. (2016). The evolution and future of manufacturing: A review. *Journal of Manufacturing Systems, 39*, 79–100.

Farrell, M.J. (1957). The measurement of productive efficiency. *Journal of the Royal Statistical Society, 120*(3), 253–290.

Halkos, G.E., & Tzeremes, N.G. (2012). Industry performance evaluation with the use of financial ratios: An application of bootstrapped DEA. *Expert Systems with Applications, 39*(5), 5872–5880.

He, F., Zhang, Q., Lei, J., Fu, W., & Xu, X. (2013). Energy efficiency and productivity change of China's iron and steel industry: Accounting for undesirable outputs. *Energy Policy, 54*, 204–213.

Jagoda, K., & Thangarajah, P. (2014). A DEA approach for improving productivity of packaging production lines: A case study. *Production Planning & Control, 25*(2), 193–202.

Jahanshahloo, G.R., Sanei, M., Rostamy-Malkhalifeh, M., & Saleh, H. (2009). A comment on A fuzzy DEA/AR approach to the selection of flexible manufacturing systems. *Computers & Industrial Engineering, 56*(4), 1713–1714.

Jain, S., Triantis, K.P., & Liu, S. (2011). Manufacturing performance measurement and target setting: A data envelopment analysis approach. *European Journal of Operational Research, 214*(3), 616–626.

Karsak, E.E. (1998). A two-phase robot selection procedure. *Production Planning & Control, 9*(7), 675–684.

Karsak, E. E., & Ahiska, S. S. (2005). Practical common weight multi-criteria decision-making approach with an improved discriminating power for technology selection. *International Journal of Production Research, 43*(8), 1537–1554.

Khouja, M. (1995). The use of data envelopment analysis for technology selection. *Computers & Industrial Engineering, 28*(1), 123–132.

Leachman, C., Pegels, C.C., & Kyoon Shin, S. (2005). Manufacturing performance: Evaluation and determinants. *International Journal of Operations & Production Management, 25*(9), 851–874.

Lee, C.Y., & Johnson, A.L. (2011). A decomposition of productivity change in the semiconductor manufacturing industry. *International Journal of Production Research, 49*(16), 4761–4785.

Lee, Z.Y., & Pai, C.C. (2011). Operation analysis and performance assessment for TFT-LCD manufacturers using improved DEA. *Expert Systems with Applications, 38*(4), 4014–4024.

Li, K., & Lin, B. (2016). Impact of energy conservation policies on the green productivity in China's manufacturing sector: Evidence from a three-stage DEA model. *Applied Energy, 168*, 351–363.

Liu, C.H., Lin, S.J., & Lewis, C. (2010). Evaluation of thermal power plant operational performance in Taiwan by data envelopment analysis. *Energy Policy, 38*(2), 1049–1058.

Liu, S.T. (2008). A fuzzy DEA/AR approach to the selection of flexible manufacturing systems. *Computers & Industrial Engineering, 54*(1), 66–76.

Lozano, S., Villa, G., & Eguía, I. (2017). Data envelopment analysis with multiple modes of functioning: Application to reconfigurable manufacturing systems. *International Journal of Production Research, 55*(24), 7566–7583.

Mahadevan, R. (2002). A DEA approach to understanding the productivity growth of Malaysia's manufacturing industries. *Asia Pacific Journal of Management, 19*(4), 587–600.

Mandal, S.K. (2010). Do undesirable output and environmental regulation matter in energy efficiency analysis? Evidence from Indian cement industry. *Energy Policy, 38*(10), 6076–6083.

Metters, R., & Vargas, V. (1999). A comparison of production scheduling policies on costs, service level, and schedule changes. *Production and Operations Management, 8*(1), 76–91.

Meeusen, W., & van Den Broeck, J. (1977). Efficiency estimation from Cobb-Douglas production functions with composed error. *International Economic Review*, 435–444.

Mitra Debnath, R., & Sebastian, V.J. (2014). Efficiency in the Indian iron and steel industry—an application of data envelopment analysis. *Journal of Advances in Management Research, 11*(1), 4–19.
Mok, V., Yeung, G., Han, Z., & Li, Z. (2007). Leverage, technical efficiency and profitability: An application of DEA to foreign-invested toy manufacturing firms in China. *Journal of Contemporary China, 16*(51), 259–274.
Mukherjee, K. (2008). Energy use efficiency in the Indian manufacturing sector: An interstate analysis. *Energy Policy, 36*(2), 662–672.
Narasimhan, R., Talluri, S., & Das, A. (2004). Exploring flexibility and execution competencies of manufacturing firms. *Journal of Operations Management, 22*(1), 91–106.
Ng, Y.C., & Chang, M.K. (2003). Impact of computerization on firm performance: A case of Shanghai manufacturing enterprises. *Journal of the Operational Research Society, 54*(10), 1029–1037.
Oggioni, G., Riccardi, R., & Toninelli, R. (2011). Eco-efficiency of the world cement industry: A data envelopment analysis. *Energy Policy, 39*(5), 2842–2854.
Önüt, S., & Soner, S. (2007). Analysis of energy use and efficiency in Turkish manufacturing sector SMEs. *Energy Conversion and Management, 48*(2), 384–394.
Park, J., Lee, D., & Zhu, J. (2014). An integrated approach for ship block manufacturing process performance evaluation: Case from a Korean shipbuilding company. *International Journal of Production Economics, 156*, 214–222.
Perroni, M.G., da Costa, S.E.G., de Lima, E.P., & da Silva, W.V. (2017). The relationship between enterprise efficiency in resource use and energy efficiency practices adoption. *International Journal of Production Economics, 190*, 108–119.
Petroni, A., & Bevilacqua, M. (2002). Identifying manufacturing flexibility best practices in small and medium enterprises. *International Journal of Operations & Production Management, 22*(8), 929–947.
Riccardi, R., Oggioni, G., & Toninelli, R. (2012). Efficiency analysis of world cement industry in presence of undesirable output: Application of data envelopment analysis and directional distance function. *Energy Policy, 44*, 140–152.
Ross, A., & Ernstberger, K. (2006). Benchmarking the IT productivity paradox: Recent evidence from the manufacturing sector. *Mathematical and Computer Modelling, 44*(1–2), 30–42.
Saaty, T.L. (2008). Decision making with the analytic hierarchy process. *International Journal of Services Sciences, 1*(1), 83–98.
Saranga, H. (2009). The Indian auto component industry—estimation of operational efficiency and its determinants using DEA. *European Journal of Operational Research, 196*(2), 707–718.
Sarkis, J. (1997). Evaluating flexible manufacturing systems alternatives using data envelopment analysis. *The Engineering Economist, 43*(1), 25–47.
Sengupta, J. K. (1992). A fuzzy systems approach in data envelopment analysis. *Computers & Mathematics with Applications, 24*(8–9), 259–266.
Shang, J., & Sueyoshi, T. (1995). A unified framework for the selection of a flexible manufacturing system. *European Journal of Operational Research, 85*(2), 297–315.
Sheu, D.D., & Peng, S.L. (2003). Assessing manufacturing management performance for notebook computer plants in Taiwan. *International Journal of Production Economics, 84*(2), 215–228.
Shrivastava, N., Sharma, S., & Chauhan, K. (2012). Efficiency assessment and benchmarking of thermal power plants in India. *Energy Policy, 40*, 159–176.

Simar, L., & Wilson, P. W. (2007). Estimation and inference in two-stage, semi-parametric models of production processes. *Journal of Econometrics, 136*(1), 31–64.

Sözen, A., Alp, I., & Özdemir, A. (2010). Assessment of operational and environmental performance of the thermal power plants in Turkey by using data envelopment analysis. *Energy Policy, 38*(10), 6194–6203.

Talluri, S., & Yoon, K.P. (2000). A cone-ratio DEA approach for AMT justification. *International Journal of Production Economics, 66*(2), 119–129.

Talluri, S., Vickery, S.K., & Droge, C.L. (2003). Transmuting performance on manufacturing dimensions into business performance: An exploratory analysis of efficiency using DEA. *International Journal of Production Research, 41*(10), 2107–2123.

Tsaur, R.C., Chen, I.F., & Chan, Y.S. (2017). TFT-LCD industry performance analysis and evaluation using GRA and DEA models. *International Journal of Production Research, 55*(15), 4378–4391.

Wang, C.H., & Chien, Y.W. (2016). Combining balanced scorecard with data envelopment analysis to conduct performance diagnosis for Taiwanese LED manufacturers. *International Journal of Production Research, 54*(17), 5169–5181.

Wang, W.K., Chan, Y.C., Lu, W.M., & Chang, H. (2015). The impacts of asset impairments on performance in the Taiwan listed electronics industry. *International Journal of Production Research, 53*(8), 2410–2426.

Wang, Y.M., & Chin, K.S. (2009). A new approach for the selection of advanced manufacturing technologies: DEA with double frontiers. *International Journal of Production Research, 47*(23), 6663–6679.

Wu, F., Fan, L.W., Zhou, P., & Zhou, D.Q. (2012). Industrial energy efficiency with CO2 emissions in China: A nonparametric analysis. *Energy Policy, 49*, 164–172.

Wu, H., Lv, K., Liang, L., & Hu, H. (2017). Measuring performance of sustainable manufacturing with recyclable wastes: A case from China's iron and steel industry. *Omega, 66*, 38–47.

Wu, J., An, Q., Xiong, B., & Chen, Y. (2013). Congestion measurement for regional industries in China: A data envelopment analysis approach with undesirable outputs. *Energy Policy, 57*, 7–13.

Wu, J., Chu, J., Zhu, Q., Yin, P., & Liang, L. (2016). DEA cross-efficiency evaluation based on satisfaction degree: An application to technology selection. *International Journal of Production Research, 54*(20), 5990–6007.

Zaim, O. (2004). Measuring environmental performance of state manufacturing through changes in pollution intensities: A DEA framework. *Ecological Economics, 48*(1), 37–47.

Zeydan, M., &Çolpan, C. (2009). A new decision support system for performance measurement using combined fuzzy TOPSIS/DEA approach. *International Journal of Production Research, 47*(15), 4327–4349.

Zhou, Z., Yang, W., Ma, C., & Liu, W. (2010). A comment on "A comment on 'A fuzzy DEA/AR approach to the selection of flexible manufacturing systems'" and "A fuzzy DEA/AR approach to the selection of flexible manufacturing systems." *Computers & Industrial Engineering, 59*(4), 1019–1021.

4

Optimization of Process Parameters for Electrical Discharge Machining of Al7075-B4C and TiC Hybrid Composite Using ELECTRE Method

M.K. Pradhan and Akash Dehari
Maulana Azad National Institute of Technology, Bhopal, MP, India

CONTENTS

4.1 Introduction 58
 4.1.1 Electrical Discharge Machining 59
 4.1.2 Mechanism of EDM Process 59
 4.1.3 Types of Electric Discharge Machines 60
 4.1.4 EDM Process Parameters 60
4.2 ELECTRE Method 62
4.3 Literature Review 63
4.4 Experimentation 64
 4.4.1 Experimental Set-up 64
 4.4.2 Machining Performance Evaluations 66
  4.4.2.1 Material Removal Rate and Tool Wear Rate 66
  4.4.2.2 Radial Overcut 66
  4.4.2.3 Surface Roughness Measurements 67
4.5 Optimization 67
 4.5.1 Optimization Process 67
 4.5.2 Multiple Attribute Decision-Making Methods 67
 4.5.3 Analytic Hierarchy Process Method 68
 4.5.4 Framing the Decision Table (Rao 2007) 68
 4.5.5 Opting Weights of the Attributes 68
 4.5.6 ELECTRE Method 71
  4.5.6.1 Formulating the Decision Table 72
  4.5.6.2 Determining Weights for the Attributes 72
  4.5.6.3 Standard Deviation Method 75
  4.5.6.4 Ranking 76

4.6 Result and Discussion 76
4.6.1 The Effect of EDM Parameters on Machining Characteristics 76
4.6.1.1 Surface Roughness (Ra) 76
4.6.2 Radial Over-Cut 77
4.6.3 Tool Wear Rate 77
4.6.4 Metal Removing Rate 78
4.6.5 Optimization of EDM Machining Parameters 79
4.7 Conclusion 79
References 80

4.1 Introduction

In this research work, an investigation has been done on the effect of EDM machining parameters on machining characteristics of EDM machine for the Al7075-B4C and Tic hybrid composite and multi-objective variable decision-making methods such as the ELECTRE optimization algorithm technique, which is used in a multi-objective problem in various fields. Optimization of EDM process parameters and multi responses has been applied by the ELECTRE algorithm. EDM is an unconventional manufacturing process, widely used in modern manufacturing industries, to make a more precise, accurate and attractive component. The composite has greater strength and toughness and is categorized as a "difficult to machine" material, which poses a significant challenge on conventional machining, however, EDM is preferred as it does not face the limitations by such materials. In this study, the main focus is to increase MRR and minimize surface roughness, tool wear rate and radial of cut for composite. The Al7075-B4C and Tic hybrid composite was selected to analyze in this optimization method. The various EDM input factors such as V, Ip, Ton and wt.% of reinforcement materials have applied to the machining of a composite. The analytic hierarchy process (AHP) approach is used to decide the weight of the relative significance of the attributes in the ELECTRE method. The improved ELECTRE (ELimination EtChoix Traduisant la REalite) method is a part of the decision-making process. ELECTRE method is used for ranking the alternatives. The ELECTRE method deals with the out-ranking relationships by using fire-wise comparison amongst the alternatives under each of the criteria discretely. By comparing among the alternatives, the preferable alternatives in the machining of hybrid composite Al7075-B4C and TIC by EDM machine are identified and declared. The Taguchi method is used for modelling and analyzes the effect of the EDM parameters on responses. The objective of the chapter is to select the best alternatives to get optimal attributes in EDM machining of a hybrid composite Al7075-B4C and Tic by using the ELECTRE method. The EDM analysis for material removal rate, tool wear rate, radial overcut, surface roughness and analyzing parameters through ELECTRE Method. The Taguchi method is used for modelling and analyzes the effect of the EDM parameters on responses. The objective of the chapter is to select the best alternatives to get optimal attributes in EDM machining of a hybrid composite (Al7075-B4C and Tic) by using the ELECTRE method.

4.1.1 Electrical Discharge Machining

The theory of electrical discharge machining, based on thermoelectric energy, uses electrical sparks for a short period of time and applies a high current density between parts and electrodes. The EDM process commonly uses thermal energy to produce heat that melts and even vaporizes the workpiece by ionization within the dielectric fluid. It produces impulsive pressure by the dielectric explosion in order to remove the melted material. Therefore, the quantity of heat produced can be controlled to make intricate and accurate components of the machines. A shaped tool is used to generate the series of spark discharges that takes place in a minor gap between the electrode and workpiece. Hence, it takes away the unwanted materials from the workpiece during vaporizing and melting. A spark is feasible as long as the tool material is conductive, and the workpiece must have electrical conduction (Pradhan and Biswas 2009). The electrical field established within the spark gap increases once the electrode travels close to the work material, and this initiates the breakdown of the dielectric fluid. Voltage decreases and drops down where the current grows rapidly right after the breakdown of the dielectric fluid, due to which the dielectric fluid is ionized, and a plasma channel grows amongst the workpiece and the electrode. The plasma channel swells up, owing to the non-stop exchange of electrons and ions (Gullu and Atici 2006). As a result of this incident, continuous heating of the work material takes place that causes a local increase of temperature on the order of 8,000 °C to 12,000 °C (Pradhan 2013). Consequently, evaporation and melting occur from both the electrodes, and then a small pool of molten metal is formed. At the end of the discharge, the current and voltage break and, therefore, the plasma tears due to the pressure exploding by the surroundings and the molten metal pool is strongly sucked up into the dielectric, producing a small crater on the surface of the workpiece—this causes melting together with evaporation. The evaporated material is eliminated from the workpiece, and by a continuous flow of dielectric fluid, melted material is cleaned out as debris.

4.1.2 Mechanism of EDM Process

During the electrical discharge machining process, the material is removed from the workpiece owing to erosion by repeated electric spark discharge occurring in the electrode and workpiece. A tiny gap is maintained between the tool and workpiece by a servo system, and both electrode and workpiece are submerged in a dielectric fluid (Pradhan et al. 2009a). The electrode generally acts as a cathode, and the workpiece acts as an anode, and a high and potential difference is applied. When the voltage is sufficiently developed through the interval, it discharges through the gap as a spark in the very small interval (one-tenth of a microsecond), the electrons and positive ions are accelerated, generating a discharge channel that becomes conductive. It is a point where the spark jumps and causes a collision between electrons and ions, creating a plasma channel. The plasma channel is no longer sustained. It produces pressure or shock waves, which preceding the molten material creates a crater of the material at the location of the spark. An instant drop in the electric resistance of the preceding spark allows current density to reach extremely high values, creating enhanced ionization and the formation of a powerful magnetic field (Pradhan and Das 2011). The instant spark takes place, and sufficiently high pressure is developed between electrode and workpiece. Such localized extreme rise in temperature leads to melting and immediate vaporization of the material, which are eroded by the flushing of the dielectric (Pradhan et al. 2009b and 2009c; Das and Pradhan 2013).

4.1.3 Types of Electric Discharge Machines

EDM facilitates the machining in a number of ways, a lot of these operations are similar to the conventional machining operation, for instance, milling and die sinking. A variety of classifications are possible and recent developments in its technology append new operations owing to increase in various requirements. A simple and general classification can be given in view of standard applications, such as:

1. Wire EDM
2. EDM milling
3. Micro-EDM
4. Electric discharge grinding (EDG)
5. Electrical discharge texture (EDT).

4.1.4 EDM Process Parameters

Some important parameters that affect the responses are the following (Puhan 2012):

Discharge voltage: It is an open circuit voltage that can be put on involving the electrodes. It deionizes the dielectric medium that depends on the resistance of the dielectric. Once the current flows, the voltage of the open circuit drops right away, and the electrode steadies the gap. It is an important parameter that controls the spark energy responsible for the higher rough surfaces, higher tool wear rate and higher MRR.

Discharge current: This is one of the most important parameters in EDM machining because it relates to power and strength during machining. This continues to grow up until it attains a predetermined level, which has been expressed as discharge current. The maximum number of amperes to use is controlled by the surface area of the cutting tools for combining the workpiece. The MRR improves at higher currents, but at the expense of surface finish and tool wear. This is an important consideration in the EDM because the precision of the machined cavity, which is a replica of the electrode device, will be affected due to excessive wear.

Pulse-on time: This is the time in which real machining happens and is assessed in µs. In every discharge cycle, there is a pause time, and the voltage applied between the electrodes occurs during the period of Ton. If there is a long pulse duration, the energy of the spark will be more intense, which makes a wide and deep crater. This is due to the fact that the removal of the material by the amount of energy applied at this time is directly proportional. Although with higher tones MRR will be higher, some surface is built through high spark energy.

Pulse-off time or pause time: In a cycle, there is a pause or a pulse off time, in which the supply of voltage is disconnected, as in effect the Ip diminishes to zero. It is as well a period of time after which the subsequent spark occurs and is expressed in µs according to a Ton. As the dielectrics must be deionized after the sparking and regain their strength, it requires some time, and furthermore, the flushing of debris too proceeds during the Toff time. The cycle is complete when enough Toff is allowed before the beginning of the next cycle. Since pressing time is a non-productive time, it should not be too small because a small amount of time makes the next spark unstable. The sum of a Ton and Toff is called pulse period or full cycle time.

Duty cycle: It is the ratio of pulse time to the pulse period and is expressed in percentage as illustrated below in equation (4.1).

$$Tau = \frac{Ton}{Ton + Toff} \times 100 \tag{4.1}$$

At high Tau, the spark energy is delivered for an extensive intermission of the pulse period follow-on in a higher machining efficiency.

Polarity: Polarity can be either straight or reverse. In straight or positive polarity the workpiece is positive, while workpiece is negative in reverse polarity. Changing the polarity can have a spectacular effect, traditionally electrode with negative polarity cut faster while positive polarity wear loss.

4.2 ELECTRE Method

The ELECTRE method is a part of the decision-making process. The ELECTRE method is nothing but elimination and choice translating reality. The ELECTRE method deals with the outranking relationships by using fire-wise comparison among the alternatives under each of the criteria separately. We can do the ELECTRE method by categorizing in three steps. The flow chart is depicted showing the steps to be followed in Figure 4.1.

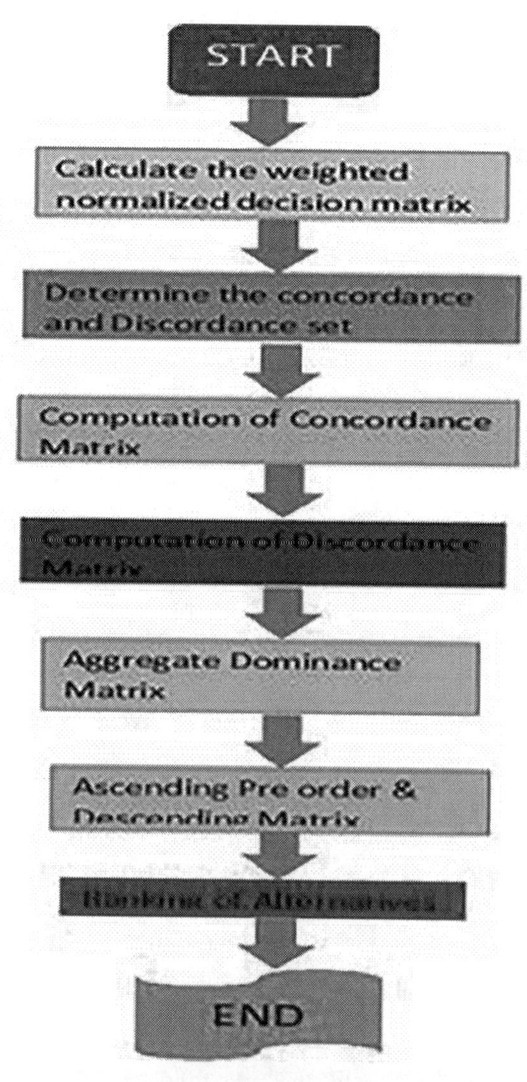

FIGURE 4.1
Flow chart for ELECTRE method

1. Normalizing the decision matrix.
2. Formation of weighting the normalizing decision matrix.
3. Determination of the concordance and discordance set.

4.3 Literature Review

A wide range of articles has been taken into account for the technique of yielding best EDM performance measures for enhancing MRR, lowering TWR and acceptable ROC. In the past, significant improvements have been made to improve production and accuracy, and therefore increase the versatility of the EDM process. The main issue is to choose the influencing machining parameters, namely Ip, Ton, Tau, V, flushing pressure, dielectric fluid and polarity in such a manner that the beneficial responses, namely MRR and accuracy increase, and at the same time ROC, TWR, and SR should make smaller to the acceptable limit. Pradhan and Biswas (2010) showed a neural network model and a couple of neuro-fuzzy models for forecasts evaluations of MRR, TWR, and ROC while machining with die-sinking EDM. The Ip, Ton, Tau and V were the inputs to the network. The Ip was the most influencing factor for MRR and G, with the maximum degree of contributions of around 87.61% and 81.90%, respectively. For TWR, Ton has the highest degree of contribution of 46.05% and is the most important factor. Dewangan and Biswas (2013) used Gray Taguchi analyzed for the optimization of EDM process parameters on AISIP20 tool steel. The Ip, Ton, work time, lift time and inter-electrode gap are considered as inputs parameters. Results revealed that MRR falls with the electrode lift time, nevertheless working time grew slightly. TWR is directly proportional to lift time and working time and inter-electrode gap has no major effect on it. The optimal parametric combination for the minimum TRR maximum MRR was obtained by Gray relational analysis. Gopalakannan et al. (2012) analyzed that the two most influential parameters that affect the MRR are Ip and Ton. MRR increases with the increase of the Ton; with the further increase of Ton the MRR starts to decrease. The value SR increases with increase in Ip and Ton. Jaharah et al. (2008) studied on responses like MRR and TWR of the tool steel. It was found that Ip was the major factor which influences MRR. Higher MRR was obtained with high Ip, medium Ton, and low Toff. However, reduced TWR was obtained at high Ip, high Ton, and at a low level of Toff. Chiang (2008) projected a mathematical model and explored the effect of Ip, Ton, Tau, voltage and their interactions on Ra. The experimentations are conducted on the Al_2O_3 + TiC workpiece and found that Ip and Ton have statistical impact on Ra. It is appealed to fit and calculate Ra narrowly with a 95% confidence interval. Kanagarajan et al. (2008) in his investigation used Ip, Ton, electrode rotation, and flushing pressure as the factor to study the Ra when machining the tungsten carbide/cobalt cemented carbide. The utmost effective parameters for reducing Ra using RSM have been acknowledged and verified experimentally. Mohri et al. (1995) investigated that TWR is affected by the precipitation of turbostratic carbon from the hydrocarbon dielectric on the electrode surface during sparking. In addition, it was also explained that due to rapid wear on the electrode edge was owing to the failure of carbon the rapid wear on the electrode edge at difficult-to-reach regions. Roy et al. (2016) used RSM to investigate the effects of the process parameters of powder mixed EDM on MRR, TWR, and Ra. Process parameters were optimized for maximize MRR, minimize TWR and Ra using the desirability function approach of MINITAB software. Owing to frequent short-circuiting, the addition of Al powder to the dielectric fluid MRR decreases, while TWR decreases for a low peak current of 2 amp with the increase in the concentration of powder. Puertas et al. (2004) analyzed the effect of EDM parameters on MRR and electrode wear in cobalt-bonded tungsten carbide workpiece. A quadratic model was developed for the responses, and it was reported that for the MRR, the Ip was the most important parameter, after that there were interaction effects of Tau, Ton and the first two. The value of MRR increases in intensity when the current and Tau were increased and decreased with Ton. Karantzalis et al. (1997) studied the mechanical properties of the Al-TiC MMC manufactured by a flux-casting technique. The general characterization of mechanical properties of Al-TIC composites contain different levels of particle addition, are calculated and studied that the 0.2% proof stress and ultimate tensile stress gradually elevated by adding up more reinforcement and the ductility diminishes firmly. When 18 vol.% of particles are added, the 0.2% proof stress and Ultimate Tensile Strength (UTS) is almost double the base alloy and you reduce the ductility for approximately a quarter of the unreinforced material. The UTS increased from 211 MPa to 465 MPa. Dwivedi et al. (2017) investigated the RHA and B4C reinforced aluminium alloy hybrid metal composite matrix fabricated by the stir casting process. The relation between the hardness of the manufactured AA6082/RHA/B4C composites with the different weight of RHA and B4C maximum toughness was observed in AA6082/7.5% B4C/2.5% RHA. The maximum toughness was found to be 23 J. After heat treatment toughness was improved, about 17.39% increases with an increase in the percentage of B4C particles up to 7.5 wt.% in AA6082/2.5 wt.% RHA.

4.4 Experimentation

4.4.1 Experimental Set-up

The experiments have been carried out to explore the influence of EDM machining parameters on machining characteristics of EDM machine

(Figure 4.2) for Al7075/B4C/TiC hybrid composite (Figure 4.3) and multi-objective variable decision-making method such as the ELECTRE optimization algorithm technique, which is used in a multi-objective problem in various fields. Optimization of EDM process parameters and multi-responses has been applied by ELECTRE algorithm. In this study, the main focus is to increase MRR and minimize surface roughness, tool wear rate

FIGURE 4.2
CNC EDM machine

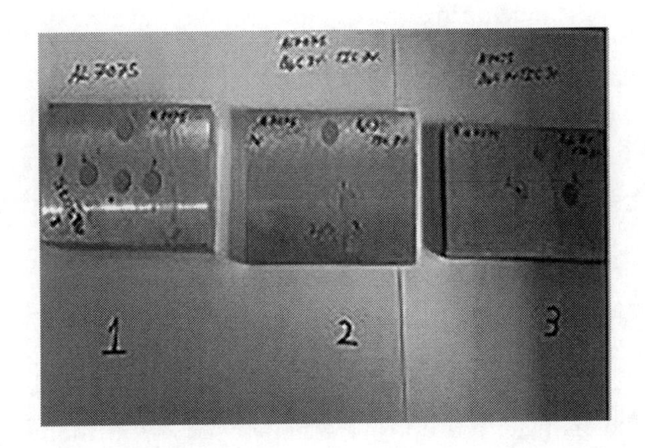

FIGURE 4.3
EDMed hybrid composite samples

and radial of cut for composite. The Al7075/B4C/TiC hybrid composite has selected to analyze in this optimization method. The various EDM input parameters such as discharge voltage, Ip, Ton and wt.% of reinforcement materials have applied to the machine of a composite. The EDM machining was done to find MRR, TWR, Ra and ROC and analyze these parameters. We can apply the ELECTRE method to get optimized EDM machining condition for composite.

4.4.2 Machining Performance Evaluations

An experimental investigation is carried out to find out the productivity, and accuracy of the EDMed surface of the hybrid composite. Parametric analysis has been carried out. The investigating factors were (V), (Ton), (Ip).

4.4.2.1 *Material Removal Rate and Tool Wear Rate*

The MRR and TWR are estimated by using the volumetric loss from the workpiece divided by the machining time (Pradhan et al. 2010). The estimated weight loss be transformed into volumetric loss as per equations 4.2 and 4.3:

$$MRR = \frac{\Delta V_w}{t} = \frac{\Delta W_w}{\rho_w t} \tag{4.2}$$

$$TWR = \frac{\Delta V_t}{t} = \frac{\Delta W_t}{\rho_w t} \tag{4.3}$$

Where ΔV_w and ΔV_t are the volume loss from the workpiece and tool, ΔW_w and ΔW_{wt} are the weight loss from the workpiece and tool respectively, t is the duration of the machining process, and ρ_w and ρ_t are the densities of the workpiece.

4.4.2.2 *Radial Overcut*

ROC (µm) is addressed as half the difference of diameter of the hole generated to the tool diameter, that is

$$ROC = \frac{d_i - d_t}{2} \tag{4.4}$$

Where d_t is the diameter tool and d_i is the diameter of the impression or cavity produced by the tool on the workpiece.

4.4.2.3 *Surface Roughness Measurements*

This is the arithmetic of the variations of the profile of the mean line.

$$Ra = \frac{1}{L}\int_0^L \left| y(x)dx \right| \tag{4.5}$$

4.5 Optimization

4.5.1 Optimization Process

To get a more accurate and optimum result of EDM machining responses on composite (Al7075 + B4C + TiC) on the EDM machine, on the basis of the multi-criteria decision-making (MCDM) process, we optimize the EDM parameters which are recorded during the time of experiment for the machining of hybrid composite. The AHP method is used to find the normalized wt. of EDM responses, and then after that we find preference function values by using optimization by the ELECTRE method to rank the alternatives.

4.5.2 Multiple Attribute Decision-Making Methods

Multiple attribute decision-making (MADM) is the process of decision-making problem-solving methods also known as the branch of decision-making problems. It is used for the operation research model that addresses the problems of decision-making in the presence of many features of decision. Each option has to choose one of the appropriate alternatives, according to the features of each attribute.

These models are often called MCDM, in the existence of many decision-makers usually criteria conflicting in nature. Based on the alternative domain, MCDM is categorized into multiple-purpose decision-making (MODM) in addition to MADM. MODM methods have values of decision variables which are set in the integer domain or in a huge number of alternative alternatives or are infinite; the best decision must be completed and satisfied and priorities must be set in the constraints. Instead, the MADM focuses on problems with discrete decision-making locations and in these problems the set of decision options has been pre-established. The various steps to decide on decision-making problem are enumerated as (Rao 2007).

1. Analytic Hierarchy Process (AHP)
2. Data envelopment analysis (DEA)
3. Preference ranking organization method for enrichment evaluations (PROMETHEE)

4. Elimination et choix traduisant la realite (ELECTRE)
5. Complex proportional assessment (COPRAS)
6. Utility additive (UTA)
7. Ordered weighted averaging (OWA) methods
8. Visekriterijumsko kompromisno rangiranje (VIKOR)
9. Gray relational analysis (GRA).

4.5.3 Analytic Hierarchy Process Method

To solve the complex decision-making problem in the analytic hierarchy process is one of the most accepted analytical techniques. The AHP hierarchy has many levels to fully characterize a particular decision problem. There are a number of functional characteristics used in the AHP method which make it a beneficial methodology. It has the capability to govern decision-making decisions, including subjective judgment, multiple decision-makers and the competence to provide stability preferences. AHP is designed for the way people really think; it is a highly regarded and widely used decision-making method. AHP cannot treat subjective characteristics together with the objective. In this method, a pairing comparison matrix is constructed using the relative importance scale. Decisions are recorded using the fundamental scale of AHP. The method determines the weight continuously and evaluates the overall performance score of the rank of options and as per the higher performance of the score. The steps of the AHP method are given below.

4.5.4 Framing the Decision Table (Rao 2007)

Step 1: First we recognize the attributes of the decision-making problem and we satisfy alternative requirements based on the identified characteristics. A quantitative or qualitative value or its limit can be allocated as a limited price or a limit value for its acceptance for the application considered as each identified attribute. An alternative with each of its attribute, meeting the requirements, may be shortlisted. The preselected options can be assessed using the suggested methodology. The value associated with the attributes for various options can be based on the existing data or the presumptions concocted by the decision-maker. EDM parameters are included and enumerated in Table 4.1 and the further selection of attributes and alternatives is shown in Table 4.2.

4.5.5 Opting Weights of the Attributes

Step 2: In relation to the purpose, the significance of the relationship is identified with different attributes. To do this, a pairing comparison matrix was established using the relative importance scale. The value that is compared to itself is always assigned to the value 1, so the main diagonal entries of

TABLE 4.1
Experimental table

Exp. No.	Ip (amp.)	Ton (μsec)	V (volts)	MRR (mm^3/min)	TWR (mm^3/min)	Ra (μm)	ROC (μm)
1	6	75	45	15.781	0.200	9.422	5.571
2	9	100	50	18.978	0.149	7.708	5.628
3	12	150	55	75.332	0.068	10.924	5.623
4	6	100	55	61.229	0.046	9.169	5.602
5	9	150	45	40.546	0.686	11.489	5.580
6	12	75	50	42.410	0.435	7.697	5.539
7	6	150	50	9.573	0.014	10.633	5.628
8	9	75	55	11.833	0.121	8.176	5.615
9	12	100	45	26.275	0.428	9.132	5.556

TABLE 4.2
Selection of attributes and alternatives

Attributes		MRR (mm^3/min)	TWR (mm^3/min)	Ra (μm)	ROC (μm)
alternative	1	15.78117	0.200169	9.422	5.571
	2	18.97784	0.149015	7.708	5.628
	3	75.33236	0.068251	10.924	5.623
	4	61.2289442	0.046228	9.169	5.6016
	5	40.5456	0.685712	11.489	5.5796
	6	42.41039	0.435473	7.697	5.539
	7	9.57271	0.014454	10.633	5.628
	8	11.83273	0.121499	8.176	5.6145
	9	26.27459	0.428489	9.132	5.5558

the coupler comparison matrix are all 1. The numbers 3, 5, 7, and 9 correspond to the verbal assessments, moderate importance, strong importance, very strong importance, and absolute importance (with 2, 4, 6, and 8 for compromise between the previous values). The scale of this relative importance used in the AHP method is summarized in the Table 4.3.

1. Find the relative normalized weight w_j of each attribute by computing the geometric mean of the *i*th row and normalizing the geometric means of rows in the comparison matrix. This can be denoted as

$$GM_j = \left\{\prod_{j=1}^{M} r_{ij}\right\}^{1/M} \quad (4.6)$$

TABLE 4.3

Preference matrix or pair-wise comparison matrix

	MRR	TWR	Ra	ROC
MRR	1.0000	4.0000	3.0000	5.0000
TWR	0.2500	1.0000	2.0000	3.0000
Ra	0.3333	0.5000	1.0000	2.0000
ROC	0.2000	0.3333	0.5000	1.0000

GM of preference matrix

$$\begin{pmatrix} & MRR & TWR & Ra & ROC & GM \\ MRR & 1.0000 & 4.0000 & 3.0000 & 5.000 & 2.7832 \\ TWR & 0.2500 & 1.0000 & 2.0000 & 3.000 & 1.1067 \\ Ra & 0.3333 & 0.5000 & 1.0000 & 2.000 & 0.7598 \\ ROC & 0.2000 & 0.3333 & 0.5000 & 1.000 & 0.4273 \\ & & & & Sum & 5.0769 \end{pmatrix} \tag{4.7}$$

And

$$w_j = GM_j / \sum \prod_{j=1}^{M} GM_j \tag{4.8}$$

$$\begin{pmatrix} & MRR & TWR & R_a & ROC & sum \\ Normalized\ wt. & 0.548197 & 0.217982 & 0.149661 & 0.08416 & 1 \end{pmatrix} \tag{4.9}$$

The geometric mean of AHP is used to achieve the normalized weight in relation to the characteristics of the current work, since its simplicity and simplicity is to realize the maximum eigenvalue and reduce the inconsistency in the decisions.

2. Compute matrix A3 and A4 such that A3 = A1 × A2 and A4 = A3 ÷ A2, where

$$A_2 = [\omega_1, \omega_2, \ldots, \omega_M]^T \tag{4.10}$$

$$\begin{bmatrix} A_2 = [0.548197, 0.217982, 0.149661, 0.08416]^T \\ A_2 = \begin{pmatrix} 0.548197 \\ 0.217982 \\ 0.149661 \\ 0.08416 \end{pmatrix} \end{bmatrix}$$

$$A_3 = \begin{pmatrix} 2.289908 \\ 0.906833 \\ 0.609686 \\ 0.341283 \end{pmatrix}$$

$$A_4 = \begin{pmatrix} 4.177163 \\ 4.160129 \\ 4.07378 \\ 4.055172 \end{pmatrix}$$

3. Calculate the maximum eigenvalue K_{max} (i.e. the average of matrix A_4).

   $K_{max} = 4.116561$

4. Calculate the consistency index CI = (K_{max} – M) ÷ (M – 1). The smaller the value of the CI, there is a deviation from the consistency of the small.

   $CI = (4.116561\text{-}4) \div (4\text{-}1)$

   $CI = 0.038854$

5. Take the random index (RI) for the number of attributes used in decision-making. Table 4.2 displays the RI values for various number of attributes. RI = 0.9.

6. Calculate the consistency ratio CR = CI/RI.

   $CR = 0.043171/0.9$

   CR = 0.043171. Usually, a CR of 0.1 or less is considered and acceptable and it reflects an informed judgment that could be attributed to the knowledge of the analyst about the problem under study. CR is less than 0.1. Thus, there is good consistency in the judgments made.

4.5.6 ELECTRE Method

The ELECTRE method is a part of the decision-making process. The ELECTRE method is nothing but elimination and choice translating reality. The ELECTRE method deals with the outranking relationships by using fire-wise comparison among the alternatives under each of the criteria distinctly. We can do the ELECTRE method by categorizing in three steps.

1. Normalizing the decision matrix
2. Weighting the normalizing decision matrix
3. Determine the concordance and discordance set.

4.5.6.1 Formulating the Decision Table

First, we recognize the attributes for making difficult decisions and choices based on recognized attributes that meet the requirements. A quantitative if not qualitative value or its rank can be allocated as the price or value of the limited extension for acceptance for the application taking into the account as classified. With each characteristic of the same, you can preselect an option that meets the conditions. The options in the short list can be assessed using the proposed method. The values associated with the tax for several options can be based on current data or estimates made by the decision-maker. The EDM parameters have been considered for the study and shown in Table 4.2, and alternatives and attributes decided from EDM output parameters are depicted in Table 4.3.

4.5.6.2 Determining Weights for the Attributes

In the ELECTRE method, there is no certain approach to specify the weight of the relative importance of the characteristics. As a result, the AHP method is adopted by importance in relation to the characteristics. The procedure for the same is as elucidated in step 2 of the improved AHP method. Various steps of the ELECTRE method are as follows:

Step 1: Calculate the normalized decision matrix. Normalization of decision matrix enumerated and shown in Table 4.4.

$$N_{ij} = \frac{R_{ij}}{\sum_{i=j}^{9} R_{ij}^{2}}, i = 1,2,3.....9; j = 1,2,3,.....9; \quad (4.11)$$

Step 2: Estimate the weighted generalized decision matrix. We assume that "W" is a diagonal matrix whose main diameter and weight are

TABLE 4.4

Normalized decision matrix

Run.	Ip	Ton	V	MRR	TWR	Ra	ROC	N_MRR	N_TWR	N_Ra	N_ROC
1	6	75	45	15.781	0.200	9.422	5.571	0.132	0.208	0.332	0.332
2	9	100	50	18.978	0.149	7.708	5.628	0.158	0.155	0.271	0.335
3	12	150	55	75.332	0.068	10.924	5.623	0.628	0.071	0.385	0.335
4	6	100	55	61.229	0.046	9.169	5.602	0.510	0.048	0.323	0.334
5	9	150	45	40.546	0.686	11.489	5.580	0.338	0.712	0.405	0.333
6	12	75	50	42.410	0.435	7.697	5.539	0.353	0.452	0.271	0.330
7	6	150	50	9.573	0.014	10.633	5.628	0.080	0.015	0.375	0.335
8	9	75	55	11.833	0.121	8.176	5.615	0.099	0.126	0.288	0.335
9	12	100	45	26.275	0.428	9.132	5.556	0.219	0.445	0.322	0.331

calculated, weighted generalized decision matrix is evaluated and depicted in Table 4.5.

$$Vij = Nij \times Wij \quad (4.12)$$

Step 3: Determine the concordance matrix using equation 4.13 to equation 4.18 from its value lies between 0 to 1. The concordance matrix is evaluated by standard deviation method by using equation 4.20 and is depicted in Table 4.6.

Step 4: Determine the Discordance matrix by using equation 4.19 to equation 4.21 and Discordance set as shown in Table 4.7.

$$Skl = \left\{ J/Nkj^{3}\ Nij \right\} (1, 2, 3, 4.........9; k = 1) \quad (4.13)$$

TABLE 4.5

Weighted normalized decision matrix

Run.	weight	0.548197	0.217982	0.149611	0.08416	Normalized weights			
1	attr.	N_MRR	N_TWR	N_Ra	N_ROC	W_MRR	W_TWR	W_Ra	W_ROC
2	alt.1	0.132	0.208	0.332	0.332	0.072	0.045	0.049	0.028
3	alt.2	0.158	0.155	0.271	0.335	0.087	0.034	0.040	0.028
4	alt.3	0.628	0.071	0.385	0.335	0.344	0.015	0.057	0.028
5	alt.4	0.510	0.048	0.323	0.334	0.280	0.010	0.048	0.028
6	alt.5	0.338	0.712	0.405	0.333	0.185	0.155	0.060	0.028
7	alt.6	0.353	0.452	0.271	0.330	0.194	0.099	0.040	0.028
8	alt.7	0.080	0.015	0.375	0.335	0.044	0.003	0.056	0.028
9	alt.8	0.099	0.126	0.288	0.335	0.054	0.027	0.043	0.028
	alt.9	0.219	0.445	0.322	0.331	0.120	0.097	0.048	0.028

TABLE 4.6

Concordance matrix

Run.	W_MRR	W_TWR	W_Ra	W_ROC	Concordance matrix			
1	0.072	0.045	0.049	0.028	0	0	0	0
2	0.087	0.034	0.040	0.028	0	1	1	0
3	0.344	0.015	0.057	0.028	0	1	0	0
4	0.280	0.010	0.048	0.028	0	1	1	0
5	0.185	0.155	0.060	0.028	0	0	0	0
6	0.194	0.099	0.040	0.028	0	0	1	1
7	0.044	0.003	0.056	0.028	1	1	0	0
8	0.054	0.027	0.043	0.028	1	1	1	0
9	0.120	0.097	0.048	0.028	0	0	1	1

TABLE 4.7
Discordance matrix

Run	W_MRR	W_TWR	W_Ra	W_ROC	Discordance matrix	Run.	W_MRR	W_TWR
1	0.072	0.045	0.049	0.028	0.000	0	0	0
2	0.087	0.034	0.040	0.028	0.015	−0.012	−0.009	0.000
3	0.344	0.015	0.057	0.028	0.272	−0.030	0.008	0.000
4	0.280	0.010	0.048	0.028	0.208	−0.035	−0.001	0.000
5	0.185	0.155	0.060	0.028	0.113	0.110	0.011	0.000
6	0.194	0.099	0.040	0.028	0.122	0.053	−0.009	0.000
7	0.044	0.003	0.056	0.028	−0.028	−0.042	0.006	0.000
8	0.054	0.027	0.043	0.028	−0.018	−0.018	−0.007	0.000
9	0.120	0.097	0.048	0.028	0.048	0.052	−0.002	0.000

When this situation is correct, then we give "1" our place and we will apply for the discourse set out in the form of the fallows.

$$D_{kl} = \{J / N_{kj} \leq N_{ij}\} \qquad (1,2,3,4,.....,9; k \neq l) \tag{4.14}$$

$$C_{kl} \geq C \tag{4.15}$$

The inception value f can be realized as an average concordance index, and the following relation could be correct:

$$C = \frac{1}{m(m-1)} \sum_{k=1\&k\neq l}^{m} \sum_{=1\&i\neq k}^{m} i = 1C_{kl} \tag{4.16}$$

Depending on the threshold value, the elements of the concordance dominance matrix F are next computed as follows:

$$F_{kl} = 1, \quad if\ C_{kl} \geq C \tag{4.17}$$

Correspondingly, the discordance dominance matrix G is expressed using a threshold value d, where d can be described as follows:

$$d = \frac{1}{m(m-1)} \sum_{k=1\&k\neq l}^{m} \sum_{i=1\&i\neq k}^{m} i = 1d_{kl} \tag{4.18}$$

$$G_{kl} = 1. \quad if\ C_{kl} \geq C \tag{4.19}$$

$$G_{kl} = 0, \quad if\ C_{kl} \leq C \tag{4.20}$$

Step 5: The elements of the aggregate dominance matrix are calculated from equation 4.17 to equation 4.18 and it is denoted by E, aggregation of dominance matrix was enumerated as shown on Table 4.8.

TABLE 4.8

Aggregate, Discordance Matrix, Aggregate and Rank

Run	Concordance matrix				Aggregate	Discordance matrix				Aggregate	Rank
1	0	0	0	0	0	0.000	0	0	0	0	9
2	0	1	1	0	2.799	0.015	−0.012	−0.009	0.000	0.015	6
3	0	1	0	0	2.683	0.272	−0.030	0.008	0.000	0.272	1
4	0	1	1	0	2.799	0.208	−0.035	−0.001	0.000	0.208	2
5	0	0	0	0	0.000	0.113	0.110	0.011	0.000	0.113	4
6	0	0	1	1	0.126	0.122	0.053	−0.009	0.000	0.122	3
7	1	1	0	0	3.673	−0.028	−0.042	0.006	0.000	0.000	7
8	1	1	1	0	3.788	−0.018	−0.018	−0.007	0.000	0.000	8
9	0	0	1	1	0.126	0.048	0.052	−0.002	0.000	0.048	5

$$E_{kl} = 1, \quad if\ d_{kl} \leq \tag{4.21}$$

$$E_{kl} = 0, \quad if\ d_{kl} \leq \tag{4.22}$$

Step 6: Eliminate the less favorable alternatives.

$$\text{ekl} = \text{fkl} \times \text{gkl} \tag{4.23}$$

From the aggregate domination matrix, partial reference of any option can be obtained. If $ek = 1$, which signifies optional gkl is preferred to alternate fkl by making use of both concordance and discordance criteria. Final aggregation was done and ranks were obtained on Table 4.9.

4.5.6.3 Standard Deviation Method

The standard deviation method is used to calculate the weight of attributes using equation 4.24.

$$W_j = \frac{\sigma_j}{\sum_{k=1}^{m} (\sigma_j)} \tag{4.24}$$

Where σ_j is the standard deviation of the normalized vector $R_j = (R_{1j}, R_{2j},..R_{3j} \ldots R_{Nj})$ in the equation, the standard deviation method calculates the objective weights of the attributes without giving any consideration to the preferences of the decision-maker.

$$R_{ij} = \frac{m_{ij}}{\left[\sum_{i=1}^{M} m_{ij}^2\right]^{\frac{1}{2}}} \tag{4.25}$$

TABLE 4.9

Rank of the alternatives

ATT.	Ip	Ton	V	MRR	TWR	Ra	ROC	RANK
ALT.1	6	75	45	15.781	0.200	9.422	5.571	9
ALT.2	9	100	50	18.978	0.149	7.708	5.628	6
ALT.3	12	150	55	75.332	0.068	10.924	5.623	1
ALT.4	6	100	55	61.229	0.046	9.169	5.602	2
ALT.5	9	150	45	40.546	0.686	11.489	5.580	4
ALT.6	12	75	50	42.410	0.435	7.697	5.539	3
ALT.7	6	150	50	9.573	0.014	10.633	5.628	7
ALT.8	9	75	55	11.833	0.121	8.176	5.615	8
ALT.9	12	100	45	26.275	0.428	9.132	5.556	5

4.5.6.4 Ranking

Table 4.9 shows the ranking of alternatives used in EDM machining. The alternative with run order 3, 4 and 6 is the best alternative to machining of hybrid composite (Al7075 + B4C + TiC) complete ranking of alternatives in EDM machining shown in the ELECTRE method, which are given above, and we can see that alternative 3 has first rank and alternative 4 has second rank and alternative 6 has third rank as shown in Table 4.9.

4.6 Result and Discussion

4.6.1 The Effect of EDM Parameters on Machining Characteristics

The procedure of ELECTRE method to establish the optimal setting of the various processing parameters on various responses of EDMed component has been investigated. The step-by-step flow chart for various calculation is depicted in Figure 1.1.

4.6.1.1 Surface Roughness (Ra)

The S/N ratio, that is signal to noise ratio, and variance (ANOVA) is used to determine the effect of EDM parameters on Ra. The graph of S/N ratio for Ra by ANOVA shows that when the wt. of reinforcement 0% to (3%B4C, 7%TiC) in the composite the SR increases then decreases up to (7%B4C, 3%TiC). When the Ip 6 to 9 decreases then the Ra will increase from 9 to 12. When Ton increases the Ra increases gradually. The voltage decreases from 45V

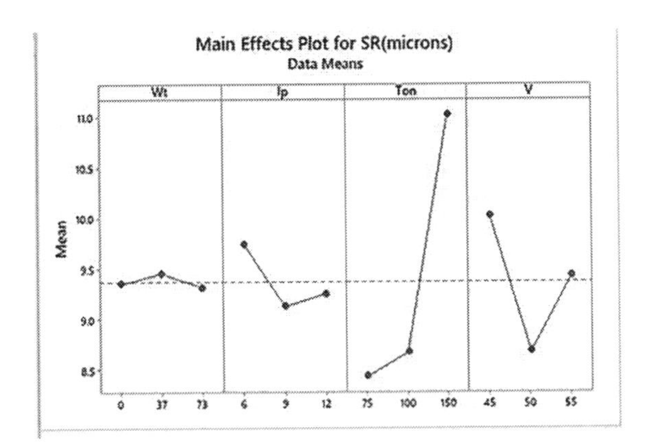

FIGURE 4.4
Schematic diagram of effect of Ra on EDM

to 50V the Ra increases after decreases up to 50V. The Graph 4 of signal to noise ratio (S/N) was plotted to see the effect of EDM parameters on surface roughness by ANOVA in Minitab software.

4.6.2 Radial Over-Cut

To regulate the influence of EDM parameters on ROC the S/N ratio and ANOVA is used. The graph of signal to noise ratio for ROC by ANOVA shows that when the wt. of reinforcement 0% to (%B4C, 7%TiC) in the composite ROC decreases then increases up to (7%B4C, 3%TiC). When Ip 6 to 9 increases then the ROC will decrease from 9 to 12. When Ton increases the ROC increases gradually and also when the voltage increases the Ra increases gradually. Figure 4.5 of the S/N ratio was plotted to see the effect of EDM parameters on radial overcut by ANOVA on Minitab software.

4.6.3 Tool Wear Rate

To know the effect of EDM parameters on TWR, the S/N ratio and ANOVA is used. The graph of signal to noise ratio(S/N) for TWR by ANOVA shows that when the wt. of reinforcement 0% to (3%B4C, 7%TiC) in the composite TWR increases then decreases up to (7%B4C, 3%TiC). When Ip 6 to 9 increases then the TWR will decrease from 9 to 12. When Ton decreases the TWR increases gradually. The voltage increases then decrease gradually. Figure 4.6 of the signal to noise ratio (S/N) was plotted to see the effect of EDM parameters on tool wear rate by ANOVA on Minitab software.

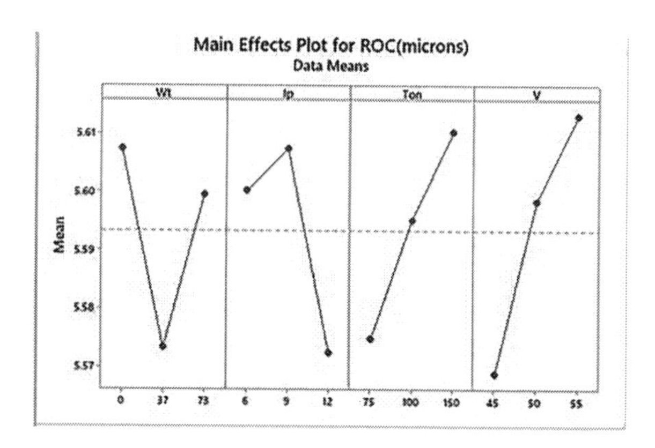

FIGURE 4.5
Schematic diagram of effect of ROC on EDM

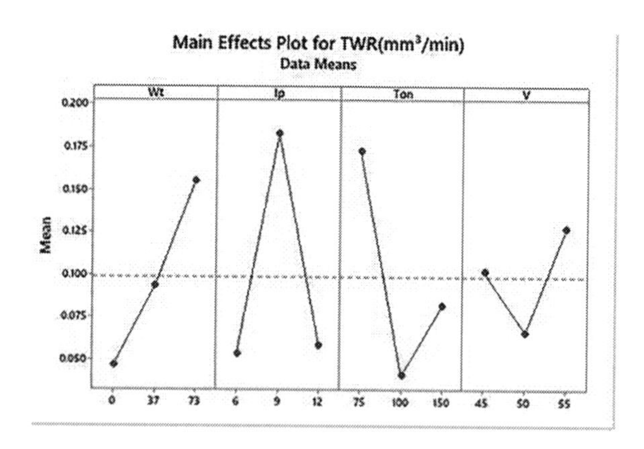

FIGURE 4.6
Schematic diagram of effect of TWR on EDM

4.6.4 Metal Removing Rate

To decide the influence of EDM parameters on MRR the S/N ratio and ANOVA is used. The graph of S/N ratio for MRR by ANOVA shows that when the wt. of reinforcement 0% to (3%B4C, 7% TiC) in the composite MRR increases then decreases up to (7%B4C, 3% TiC). When the Ip 6 to 9 decreases, the MRR will increase from 9 to 12. When Ton increases the MRR increases gradually, the voltage decreases from 45V to 50V the MRR decreases up to 50V. Figure 4.7 of the S/N ratio was plotted to see the effect of EDM parameters on metal removing rate by ANOVA on Minitab software.

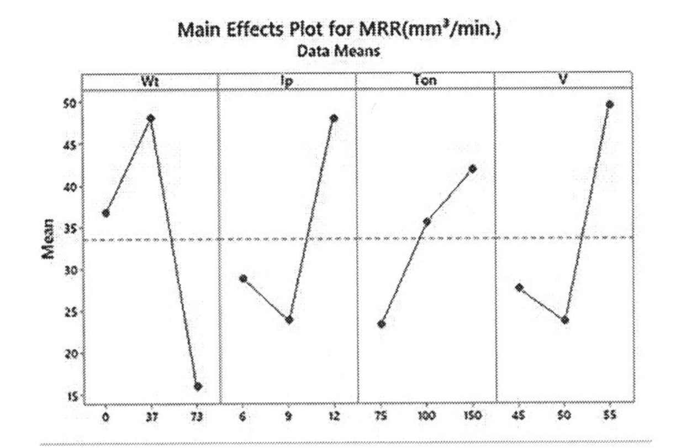

FIGURE 4.7
Schematic diagram of effect of MRR on EDM

4.6.5 Optimization of EDM Machining Parameters

To get the optimum result of the machining parameters and to reduce the manufacturing cost of the composite product. The AHP and ELECTRE method was used for the optimization of EDM parameters for machining of hybrid composite (Al7075 + B4C + TiC) and find the rank of alternatives. Alternative 3 has ranked first, alternative 4 has ranked second and alternative 6 has ranked third, so the result shows that alternatives 3, 4 and 6 are the best suitable alternative to EDM machining of hybrid composite.

4.7 Conclusion

In this investigation, the evaluation of various properties and the machinability of hybrid composite (Al7075 + 3% B4C + 7% TiC) in the EDM machine and the influence of the machining parameters have been carried out. The experiments were conducted to identify the influencing parameters Ip, Ton, V, Tau on the responses, namely MRR, TWR, Ra, and ROC. The ELECTRE method has been used to find the best machining parameters for machining of hybrid composite (Al7075 + 3% B4C + 7% TiC). It has been observed that the Ip is the most effective parameter as compared to voltage on the EDM machining characteristics. The Minitab 18 software was used for the Taguchi method to model and analyze the effect of EDM machining parameters on the responses. Finally, the soft computing techniques were employed for modelling of MRR, TWR, Ra, and Radial Over-Cut. In the optimization of EDM parameters by using the AHP method, we calculated the normalized weight for attributes such as MRR, TWR, ROC, and Ra and the ELECTRE method, the rank is given to the alternatives. It is concluded that alternative 3 is rated first, alternate 4 is rated second and alternate 3 has nine alternatives out of the third rank, so we can say that alternatives 3, 4 and 6 are the hybrid composite of the machine in most preferred alternatives (Al7075 + B4C + TIC) by the EDM machine.

References

Chiang, K. (2008). Modelling and analysis of the effects of machining parameters on the performance characteristics in the EDM process of Al2O 3 + TiC mixed ceramic. *International Journal of Advanced Manufacturing Technology, 37*(5–6), 523–533.

Das, R., & Pradhan, M.K. (2013). ANN modelling for surface roughness in electrical discharge machining: A comparative study. *International Journal of Service and Computing Oriented Manufacturing, 1*(2), 124–140.

Dewangan, S., & Biswas, C.K. (2013). Optimisation of machining parameters using grey relation analysis for EDM with impulse flushing. *International Journal of Mechatronics and Manufacturing Systems, 6*(2), 144–158.

Dwivedi, S.P., Srivastava, A., Kumar, A., & Nandan, B. (2017). Microstructure and mechanical behaviour of RHA and B4C reinforced aluminium alloy hybrid metal matrix composite. *Indian Journal of Engineering and Materials Sciences 24*, 133–140.

Gopalakannan, S., Senthilvelan, T., & Ranganathan, S. (2012). Modelling and optimization of EDM process parameters on machining of al 7075-b4c MMC using RSM. *Procedia Engineering, 38*, 685–690.

Gullu, A., & Atici, U. (2006). Investigation of the effects of plasma arc parameters on the structure variation of AISI 304 and St 52 steels. *Materials and Design, 27*, 1157–1162.

Jaharah, A.G., Liang, C.G., Wahid, S.Z., Rahman, M.N.A., & Hassan, C.H.C. (2008). Performance of copper electrode in electrical discharge machining of AISI H13 harden steel. *International Journal of Mechanical and Materials Engineering, 3*(1), 25–29.

Kanagarajan, D., Karthikeyan, R., Palanikumar, K., & Sivaraj, P. (2008). Influence of process parameters on electric-discharge machining of WC/30%Co composites. *Proceedings of the Institution of Mechanical Engineers, Part B: Journal of Engineering Manufacture, 222*(7), 807–815.

Karantzalis, A., Wyatt, S., & Kennedy, A. (1997). The mechanical properties of al-tic metal matrix composites fabricated by a flux-casting technique. *Materials Science and Engineering: A, 237*(2), 200–206.

Mohri, N., Suzuki, M., Furuya, M., Saito, N., & Kobayashi, A. (1995). Electrode wear process in electrical discharge machining. *CIRP Annals—Manufacturing Technology, 44*(1), 165–168.

Pradhan, M.K. (2013). Optimization of MRR, TWR and surface roughness of EDMed D2 Steel using an integrated approach of RSM, GRA and entropy measurement method. *2013 International Conference on Energy Efficient Technologies for Sustainability (ICEETS)*, April, IEEE, pp. 865–869.

Pradhan, M.K., & Biswas, C.K. (2009). Neuro-fuzzy model and regression model a comparison study of MRR in electrical discharge machining of D2 tool steel. *International Journal of Engineering and Applied Sciences, World Academy of Science Engineering and Technology, 5*, 328–333.

Pradhan, M.K., & Biswas, C.K. (2010). Neuro-fuzzy and Neural network-based prediction of various responses in electrical discharge machining of AISI D2 Steel. *International Journal of Advance Manufacturing Technology*. Springer, *50*, 591–610.

Pradhan, M.K., & Das, R. (2011). Recurrent neural network estimation of material removal rate in electrical discharge machining of AISI D2 tool steel. *Proceedings of the Institution of Mechanical Engineers, Part B: Journal of Engineering Manufacture, 225*(3), 414–421.

Pradhan, M.K., Das, R., & Biswas, C.K. (2009a). Comparisons of neural network models on surface roughness in electrical discharge machining. *Proceedings of the Institution of Mechanical Engineers, Part B: The Journal of Engineering Manufacture, 223*(7), 801–808.

Pradhan, M.K., Das, R., & Biswas, C.K. (2009b). Prediction of surface roughness in electrical discharge machining of D2 steel using regression and artificial neural networks modeling. *Journal of Machining and Forming Technologies (JoMFT)*. Nova Science Publishers, 2(1–2), 25–46.

Pradhan, M.K., Das, R., & Biswas, C.K. (2009c). Predictive modeling and analysis of surface roughness in electro-discharge machining of D2 tool steel using regression and neural networks approach. *International Journal of Design and Manufacturing Technologies, 3*(2).

Pradhan, M.K., Das, R., & Biswas, C.K. (2010). Prediction of material removal rate using recurrent elman networks in electrical discharge machining of AISI D2 tool steel. *International Journal of Manufacturing Technology and Industrial Engineering (IJMTIE), 1*(1), 29–37.

Pradhan, M., & Biswas, C. (2010). Neuro-fuzzy and neural network-based prediction of various responses in electrical discharge machining of AISI D2 steel—NF and ANN based prediction of responses in EDM of D2 steel. *International Journal of Advanced Manufacturing Technology, 50*, 591–610.

Puertas, I., Luis, C.J., & Alvarez, L. (2004). Analysis of the influence of EDM parameters on surface quality, MRR and EWR of WC-Co. *Journal of Materials Processing Technology, 153–154*(1–3), 1026–1032.

Puhan, D. (2012). Non-conventional machining of Al/SiC metal matrix composite. *Diss.*

Rao, R.V. (2007). *Decision making in the manufacturing environment: Using graph theory and fuzzy multiple attribute decision making methods*. London: Springer-Verlag.

Roy, C., Syed, K.H., & Kuppan, P. (2016). Machinability of al/10% sic/2.5% tib2metal matrix composite with powder-mixed electrical discharge machining. *Procedia Technology, 25*, 1056–1063.

5

Selection of Laser Micro-drilling Process Parameters Using Novel Bat Algorithm and Bird Swarm Algorithm

Bappa Acherjee

Department of Production Engineering, BIT Mesra, Ranchi – 835215, India
bappa.rana@gmail.com

Debanjan Maity

Department of Mechanical Engineering, IIT Kharagpur, Kharagpur – 721302, India

Deval Karia

Center for Product Design and Manufacturing, IISC Bangalore, Bengaluru – 560012, India

Arunanshu S. Kuar

Department of Production Engineering, Jadavpur University, Kolkata – 700032, India

CONTENTS

5.1 Introduction 84
5.2 Methodology 85
  5.2.1 Response Surface Method 86
  5.2.2 Novel Bat Algorithm 86
    5.2.2.1 Selection of Habitat 87
    5.2.2.2 Bats with Quantum Behaviour 87
    5.2.2.3 Bats with Mechanical Behaviour 88
    5.2.2.4 Local Search 88
  5.2.3 Bird Swarm Algorithm 89
    5.2.3.1 Foraging Behaviour 89
    5.2.3.2 Vigilance Behaviour 90
    5.2.3.3 Flight Behaviour 90
5.3 Experimental Results and RSM Models 91
5.4 Optimization Using NBA and BSA: Results and Discussions 92
  5.4.1 Single Objective Optimization 92
  5.4.2 Multi-objective Optimization 95
5.5 Prediction of Parametric Trends Using NBA and BSA 96
5.6 Conclusion 98
References 99

5.1 Introduction

Laser micro-drilling is a process where the material surface is gradually ablated by the exposure of a laser beam to create a blind or through hole. Laser beam materials processing has several advantages as the focused laser beam acts as a non-contact tool and thus can remove materials from hard-to-machine materials, depending on the intensity of the beam. The laser beam can be precisely controlled by using lenses to focus the laser beam to a very small spot to create micro-features. Laser systems are flexible and can be easily integrated and automated. The laser beam removes the materials from the workpieces by photo-chemical and photo-thermal actions. In photo-chemical action, the photon energy of the laser beam breaks the atomic bonds of the materials of the irradiated zone and allowing them to be expelled from the targeted zone. In photo-thermal action, the photon energy of the laser beam is absorbed by the work material and is converted to heat, which results in a temperature rise within the material. The material is removed by the action of melt ejection or by boiling and evaporation.

Despite the present level of acceptance in the industries, several defects such as spatter, recast, heat-affected zone (HAZ) and taper limit its application. Elimination of these defects is the subject of intense research. Several research works are carried out to minimize the laser micro-drilling defects and to optimize the process parameters (Yilbas and Yilbas 1987; Ghoreishi et al. 2002; Jackson and O'Neill 2003; Kuar et al. 2006; Kuar et al. 2012; Goel and Pandey 2014; Biswas et al. 2015). Previous mathematical procedures to optimize laser micro-drilling parameters have largely resulted in nearly optimal or sub-optimal results.

The novel bat algorithm (NBA) and bird swarm algorithm (BSA) are two newly developed nature-inspired metaheuristic optimization algorithms capable of solving local as well as global optimization problems. The bat algorithm (BA) is introduced by Yang (2010), especially based on echolocation behaviour of bats when searching their prey. The bat algorithm finds its application in a wide variety of optimization problems, which includes image processing (Yang 2013), scheduling (Du and Liu 2012), power system stabilizing (Musikapun and Pongcharoen 2012) and several other optimization problems (Sambariya and Prasad 2014). Meng et al. (2015) improves the bat algorithm by incorporating the bats' habitat selection and their self-adaptive compensation for the Doppler effect in echoes into the basic BA. The modified algorithm was proposed in 2015, which was termed the novel bat algorithm (NBA). When compared to the basic BA algorithm by using some benchmark problems and a number of real-world design problems, the NBA algorithm is found to be more effective and efficient than its basic version. Gautham and Rajamohan (2016) applied novel bat algorithm to solve the economic load dispatch (ELD) problem for obtaining the most efficient operation out of a power system generation network to minimize the generator fuel cost. Liu (2016) employed novel bat algorithm to answer the multi-criteria scheduling problem like a timetable problem to ensure the teachers and the students should be assigned in the appropriate time section to the appropriate classroom. The bird swarm algorithm, which is inspired by social behaviours and social interactions in bird swarms, is proposed by Meng et al. (2015, 2016). The optimization algorithm includes foraging, vigilance and flight behaviours of birds, which is essential for greater possibilities of birds' survival. Four search strategies are formulated and correlated through five basic BSA rules based on the social behaviours of the birds. The effectiveness, stability and superiority of BSA is checked and validated by using benchmark problems. Parashar et al. (2017) employed BSA for solution of various mathematical functions. A comparison is made between the results of BSA and the results of some other nature-inspired algorithms available in literature which proved the superiority of BSA over others. NBA and BSA both are developed in recent years (Meng et al. 2015, 2016) and almost unexplored non-conventional optimization techniques. Only a very few literatures are available on the applications of these two algorithms (Gautham and Rajamohan 2016; Liu 2016; Meng et al. 2016; Parashar et al. 2017) in the field of engineering. Application of these two algorithms for solving multi-criteria optimization problems in manufacturing engineering would lead to the solutions of many process optimization problems. Comparison of their effectiveness in terms of convergence speed, optimized populations, solution time and optimal results to choose the best among these two is also a topic of interest.

In this work, the novel bat algorithm and bird swarm algorithm are used for optimization of laser micro-drilling parameters. Compared to the conventional optimization technique, the optimization performance of these two algorithms are investigated. Both algorithms are compared for their accuracy, repeatability, convergence rate and computational time. The present study also reveals an analysis of trends of the responses for different control parameters.

5.2 Methodology

The response surface method (RSM) is used to establish the empirical equations to correlate the laser micro-drilling input parameters with output responses. The developed equations are further used as objective functions for determination of the optimal parameters setting using the novel bat algorithm and bird swarm algorithm. In this section, the methodologies employed for modelling, analysis and optimization of the laser micro-drilling process parameters are described in brief.

5.2.1 Response Surface Method

The response surface method is used for developing the correlation between input variables and measured responses to develop predictive models, which can further be used for optimization of the responses. In this method, the models are developed using a set of mathematical techniques, and the developed models are tested by using some statistical techniques to check their adequacies (Montgomery 2001; Khuri and Cornell 1996). These mathematical models are used to predict the values of responses for different values of a set of input parameters. Response surface y is a function of a number of independently controllable and measurable input variables ($\xi_1, \xi_2, \ldots, \xi_k$).

$$y = f\left(\xi_1, \xi_2, \ldots, \xi_k\right) + \varepsilon \tag{5.1}$$

where f is the response function and ε is experimental error due to noise which introduces variability that is not accounted for in f. The nature of function f is unknown and it depends on the experimenter's ability to approximation. Generally, a polynomial function of the first or second order is used for the models. The first-order model is used for a relatively small region of process variables to approximate the response function with little curvature. However, a second order polynomial model is used for a wider design space where a first-order model is inadequate to approximate the degree of nonlinearity.

Second-order models are most commonly used for approximation of the response function in RSM. The least squares method is used to determine the values coefficients of the function variables.

In mathematical terms, the model of the first order is expressed as:

$$\eta = \beta_0 + \beta_1 x_1 + \beta_2 x_2 + \ldots + \beta_k x_k; \tag{5.2}$$

And the model of the second order is expressed as:

$$\eta = \beta_0 + \sum_{j=1}^{k} \beta_j x_j + \sum_{j=1}^{k} \beta_{jj} x_j^2 + \sum_{i<} \sum_{j=2}^{k} \beta_{ij} x_i x_j; \tag{5.3}$$

In some infrequent situations, approximating polynomials of orders greater than two are used. The βs are a set of unknown parameters, called regression coefficients.

5.2.2 Novel Bat Algorithm

The proposed algorithm is mainly based on the habitat selection behaviour of bats and their self-adaptive compensation for the Doppler effect in echoes. An adaptive search strategy in local domain and bats' echolocation characteristics are also implemented into NBA. It has been proved that the novel bat algorithm outperforms the basic bat algorithm (Meng et al. 2015). The Doppler effect and the foraging habitats of bats are not incorporated in basic BA. However, these are taken into consideration in NBA. For mathematical formulation of NBA, two more idealized rules are added with those assumed in original BA (Yang 2010). These are:

1. All bats can forage in various habitats, contingent upon a stochastic determination
2. All bats can compensate for the Doppler effect in echoes. They can modify the compensation rate adaptively, in extent to the proximity of their objectives.

5.2.2.1 Selection of Habitat

The selection of habitats by bats is governed by several unsystematic phenomena. Thus, it is replicated as a stochastic choice. $P\in[0,1]$ corresponds to the habitat selection threshold. If R [0,1] is a random number that is uniform and smaller than P, then bats follow the quantum behaviour to look for habitats in a wide range. If not, then bats follow the mechanical behaviour to look for habitats in limited range.

5.2.2.2 Bats with Quantum Behaviour

Bats with quantum behaviour search for habitats in a wide range. When a bat locates a food site, other bats would follow it right away for food. Therefore, the global best position g_j^t served as the attractor in the bats swarm. Virtual bats' positions with quantum behaviour are expressed using the following formulae:

$$x_{i,j}^{t+1} \begin{cases} g_j^t + \theta_* \left| mean_j^t - x_{i,j}^t \right| * \ln\left(\dfrac{1}{u_{i,j}}\right), if\ r\ and_j(0,1) < 0. \\ g_j^t - \theta_* \left| mean_j^t - x_{i,j}^t \right| * \ln\left(\dfrac{1}{u_{i,j}}\right), otherwise \end{cases} \tag{5.4}$$

where, $rand(0,1)$ is a random number (Yang 2010). $x_{i,j}^{t+1}$ is the jth ($1\leq j\leq N$) component of the ith ($1\leq i\leq M$) position of the individual at step t. $u_{i,j}$ is a number that is distributed uniformly between 0 and 1. θ is known as coefficient of contraction-expansion, and

$$mean_j^t = \frac{1}{N} * \sum_i^N x_{i,j}^t \tag{5.5}$$

where, N is the number of dimensional Hilbert spaces and M is the number of individual searches.

5.2.2.3 Bats with Mechanical Behaviour

The formulae used of updating the offspring $x_{i,j}^{t+1}$ and $v_{i,j}^{t+1}$ at time step t are modified to incorporate the Doppler effect, and are different from formulae used in basic BA. Initially, each bat is randomly assigned with a frequency in between f_{min} and f_{max}, which relies upon the Doppler effect. The inertia weight, w, is utilized to update velocity. The numerical conditions are confined as takes after:

$$f_{i,j} = f_{min} + (f_{max} - f_{min}) * rand(0,1) \tag{5.6}$$

$$f_{i,j} = \frac{\left(c + v_{i,j}^{t}\right)}{\left(c + v_{g,j}^{t}\right)} * f_{i,j} * \left(1 + c_i * \frac{g_j^t - x_{i,j}^t}{\left|g_j^t - x_{i,j}^t\right| + \varepsilon}\right) \tag{5.7}$$

$$v_{i,j}^{t+1} = w * v_{i,j}^{t} + \left(g_j^t - x_{i,j}^t\right) * f_{i,j} \tag{5.8}$$

$$x_{i,j}^{t+1} = x_{i,j}^{t} + v_{i,j}^{t} \tag{5.9}$$

where, $w \in [0,1]$ is a uniform random vector, ε is a constant having very small numeric value used in computer to keep away from zero-division-error, $c \in [0,1]$ is a positive number, and $v_{g,j}^{t}$ is the speed coordinator for the best position in the global environment.

5.2.2.4 Local Search

Bats would increase the pulse emission rate and reduce the loudness. The new location of each bat is determined locally by:

$$\text{If } (r\,\text{and}(0,1) > r_i) \tag{5.10}$$

$$x_{i,j}^{t+1} = g_j^{t} * \left(1 + r\,\text{and}\,n\left(0,\sigma^2\right)\right) \tag{5.11}$$

$$\sigma^2 = \left|A_i^t - A_{mean}^t\right| + \varepsilon \tag{5.12}$$

where, r and $n(0,\sigma^2)$ take after a Gaussian distribution with standard deviation = σ^2 and mean = 0 and ε is utilized to affirm $\sigma^2 > 0$, A_{mean}^{t} is the average loudness of all the bats at time step t. r_i is the rate of pulse emission.

5.2.3 Bird Swarm Algorithm

Bird swarm algorithm is a nature inspired metaheuristic algorithm which is inspired by the social behaviour and interactions among bird swarms Meng et al. (2016). Four simplified rules assumed to define the social behaviour of the birds are as follows:

1. Every bird used to forage for food and keeps vigilance on surrounding for safety. They decide their activity to switch between foraging and vigilance behaviour as per requirements, which follow stochastic decisions.
2. During foraging the bird can recall the past best experience of food patch which they can compare and update with the present experience. The social information is also shared between the group of bird swarm.
3. Each bird tries to move to the centre of the swarm during vigilance activity but faces interference due to competition between swarm. The birds having higher reserves have greater possibility to lie within the centre of the swarm than others.
4. Birds often fly to alternative location. Birds can regularly switch between producing and scrounging when traveling to another site. The bird with the most amazing reserves is a producer, while the bird with the least reserve is a scrounger. The producer and scrounger would be randomly different birds with reserves between the highest and the least. Producers scan food effectively. Producers effectively scan for food. Scroungers would randomly search for food after a producer.

All N virtual birds, represented in step t by their position $x_i^t\left(i \in [1,2,\ldots.N]\right)$, seek food and fly in a space of D-dimensions.

5.2.3.1 Foraging Behaviour

Each bird scans food through their experience and understanding of the swarms. Rule 2 can be composed in numerical form following:

$$x_{i,j}^{t+1} = x_{i,j}^t + \left(p_{i,j} - x_{i,j}^t * c * rand(0,1) + (g_j - x_{i,j}^t\right) * s * rand(0,1)) \tag{5.13}$$

Where $j \in [1, \ldots., D]$, $rand(0,1)$ signifies independent uniformly distributed numbers in (0, 1). c and s are cognitive coefficients that accelerate socially. $p_{i,j}$ is the ith bird's best past location and g_j is the swarm's best past location. If the uniform random number in (0,1) is lower than the constant value, P ($P \in 0,1$), the bird would be the food forage. Else, the bird would be vigilant.

5.2.3.2 Vigilance Behaviour

In accordance with Rule 3, birds would try to move to the centre of the swarm and compete. Each bird would therefore not move directly to the centre of the swarm These movements can be detailed as follows:

$$x_{i,j}^{t+1} = x_{i,j}^{t} + A_1\left(mean_j - x_{i,j}^{t}\right) * rand(0,1) + A_2\left(p_{k,j} - x_{i,j}^{t}\right) * rand(-1,1) \tag{5.14}$$

$$A_1 = a_1 * \exp\left(-\frac{pF_{it_i}}{sumFit + \varepsilon} * N\right) \tag{5.15}$$

$$A_2 = a_2 * \exp\left(\left(\frac{pF_{it_i} - pF_{it_k}}{\left|pF_{it_k} - pF_{it_i}\right| + \varepsilon}\right)\frac{N * pF_{it_k}}{sumFit + \varepsilon}\right) \tag{5.16}$$

where $k(k \neq i)$ is a positive integer, selected randomly from 1 and N. a_1 and a_2 in [0,2] are positive constants. pF_{it_i} means the best fitness value of the ith bird, and *sumFit* is the sum of the best fitness value of the swarms. ε is a small constant value to stay away from the error of zero division. $mean_j$ shows the jth component of the whole swarm's average position. Since each bird must remain in the centre of the swarm, the A_1 and *rand*(0,1) product should not exceed 1. In this case, A_2 was used to reproduce the immediate impact of an interference when a bird moves to the swarm centre. If the best fitness value of a random kth bird ($k \neq i$) is higher than that of the ith bird, then $A_2 > A_2$, which means that the ith bird can be obstructed more prominently than the kth bird.

5.2.3.3 Flight Behaviour

In accordance with Rule 4, producers and scroungers can be isolated from the swarm. The practices of the producers and scroungers can be shown numerically as follows:

$$x_{i,j}^{t+1} = x_{i,j}^{t} + randn(0,1) * x_{i,j}^{t} \tag{5.17}$$

$$x_{i,j}^{t+1} = x_{i,j}^{t} + (x_{k,j}^{t} - x_{i,j}^{t}) * FL * rand(0,1) \tag{5.18}$$

where *randn*(0,1) indicates Gaussian random number with a mean = 0 and standard deviation = 1, $k \in 1, 2, 3. . ., N]$, $k \neq i$. $FL(FL \in [0,2])$ implies that the scrounger would search for food after the producer. For simplicity, we accept that every bird flies every FQ unit interval somewhere else. FQ is a positive integer here.

5.3 Experimental Results and RSM Models

A four factors five levels central composite rotatable design of response surface methodology (RSM) is utilized to design the experimental matrix and to build the mathematical models. A computer numeric controlled Nd:YAG laser is employed for micro-drilling experiments. Alumina is used as workpiece material. Table 5.1 furnishes micro-drilling parameters and their levels. Each parameter is limited to values corresponding to coded levels of –2 and + 2. Hole taper (Y_{HT}) and the HAZ width (Y_{HAZ}) are selected as performance characteristics to measure the drilled hole quality.

Design-expert software is used for analysis of the measured responses and determining the mathematical models with best fits using RSM. Backward elimination regression method has been used to eliminate the unimportant terms in the response equations. The empirical equations developed in terms of coded factors using RSM, are given below:

$$\begin{aligned} Y_{HT}(rad) = {} & 0.054663 + 0.001975A - 0.00382B \\ & + 0.000567C - 0.00101D + 0.00105C\,D \\ & + 0.000892B^2 - 0.0013D^2 \end{aligned} \quad (5.19)$$

$$\begin{aligned} Y_{HAZW}(mm) = {} & 0.323256 + 0.025438A - 0.00236B \\ & + 0.003271C - 0.00425D - 0.00601B\,D \\ & - 0.00378A^2 + 0.003692B^2 - 0.02233D^2 \end{aligned} \quad (5.20)$$

TABLE 5.1

Laser micro-drilling parameters with their coded levels (Kuar et al. 2012)

Parameters	Unit	Symbol	Levels				
			–2	–1	0	1	2
Lamp current	A	A	20	21	22	23	24
Pulse frequency	kHz	B	1	2	3	4	5
Air pressure	kg/cm^2	C	0.6	1	1.4	1.8	2.2
Pulse width	mm	D	2	6	10	14	18

5.4 Optimization Using NBA and BSA: Results and Discussions

To obtain superior micro-drill quality, it is desired that hole taper and HAZ width should be minimum. Thus, both the responses have been minimized for optimization purpose. To find a solution for single—objective optimization and multi—objective optimization of laser micro—drilling process parameters, the novel bat algorithm and bird swarm algorithm are used. For this purpose, MATLAB computer programs are developed on an Intel Core i3–380M CPU @ 2.53 GHz, 3.00GB RAM operating platform using novel bat algorithm and bird swarm algorithm. The algorithm specific parameters used are given in Table 5.2. It is observed that both the algorithms have obtained optimum values for all the test functions.

5.4.1 Single Objective Optimization

The minimum hole taper obtained by NBA and BSA is 0.0341 radian, whereas using conventional RSM-based optimization method it is obtained as 0.0465 radian. The minimum values of HAZ width obtained by NBA and BSA are 0.1389 mm; whereas using RSM-based optimization method it is obtained as 0.2596 mm. Table 5.3 presents the results, along with the optimal parameter

TABLE 5.2

Algorithm specific parameters considered for NBA

Novel bat algorithm	Parameters from bat algorithm (BA) Maximum generations (iterations) $(M) = 500$ Population size$(n) = 20$ The maximal and minimal pulse rate $(r_{0max}, r_{0min}) = 1,0$ The maximal and minimal frequency $(f_{max}, f_{min}) = 1.5,0$ The maximal and minimal loudness $(A_{max}, A_{min}) = 2,1$	New parameters (NBA) The frequency of the update of the loudness and pulse emission rate $(G) = 10$ Maximum and minimum probability of habitat selection $(P_{max}, P_{min}) = 0.9,0.6$ Maximum and minimum compensation rate for Doppler Effect in echoes $(c_{max}, c_{min}) = 0.9,0.1$ Maximum and minimum contraction/expansion coefficient $(\theta_{max}, \theta_{min}) = 1,0.5$ Maximum and minimum inertia weight $(w_{max}, w_{min}) = 0.9,0.5$
Bird swarm algorithm	Maximum generations (iterations)(M) = 500 Population size $(n) = 20$ The frequency of birds' flight behaviours $(FQ) = 10$ Cognitive accelerated coefficient (C1) = 1.5 Social accelerated coefficient (C2) = 1.5 Two parameters which are related to the indirect and direct effect on the birds' vigilance behaviours $(A1, A2) = 1,1$	

settings for the responses. The results obtained using NBA and BSA show drastic improvements, 26.6% for hole taper, and 46.5% for HAZ width, when compared to the results obtained using RSM. This drastic improvement is due to the fact that RSM performs local optimization only, whereas NBA and BSA perform global optimization. NBA and BSA give the same results as the optimized values are extreme and any further improvement in the optimized results is not possible.

Figure 5.1 shows the histograms of algorithm convergence data for NBA and BSA with respect to (a) hole taper, and (b) HAZ width. It is evident from Figure 5.1 that both the algorithms show a fast convergence to its global optima, whereas BSA shows faster convergence than NBA. It is also seen that the mean values of the results obtained by BSA (0.0347 radian for hole taper, and 0.1438 mm for HAZ width) are closer to the optimum results (0.0341 radian for hole taper, and 0.1389 mm for HAZ width), compared to the mean values obtained by NBA (0.0348 radian for hole taper, and 0.1468 mm for HAZ width). As the population mean value of the functional evaluations for BSA is much closer to its optima than that for NBA, it can be concluded that more number of evaluations have converged to its optima for BSA as compared to NBA, which is evident from Figure 5.1.

Table 5.4 furnishes the details of algorithm convergence in terms of convergence speed and number of optimized populations. During optimization of hole taper, the NBA converges to the optima after 185 evaluations, whereas only 87 evaluations have been taken by BSA. For NBA almost 7,784 evaluations have been found converged to its optimal value, whereas, for BSA it is 9,470 evaluations, as indicated by the spikes in Figure 5.1. During optimization of HAZ width, NBA takes 549 evaluations to reach the optima, whereas BSA takes only 63 evaluations. For NBA it is found only 6,724 evaluations where for BSA 9,485 evaluations have been converged to its optimal value. It is observed that BSA provides more effective and better result in terms of both convergence speed and number of optimized populations. The average

TABLE 5.3

Results of single objective optimization

Response	Nature of optimization	Optimization technique	Optimal value	Optimal parameter setting			
				A	*B*	*C*	*D*
Y_{HT}	Minimize	NBA	0.0341	20.00	5.00	0.60	18.00
		BSA	0.0341	20.00	5.00	0.60	18.00
		RSM	0.0465	21.02	4.00	1.14	14.00
Y_{HAZ}	Minimize	NBA	0.1389	20.00	4.95	0.60	18.00
		BSA	0.1389	20.00	4.95	0.60	18.00
		RSM	0.2596	21.00	4.00	1.00	13.99

(a)

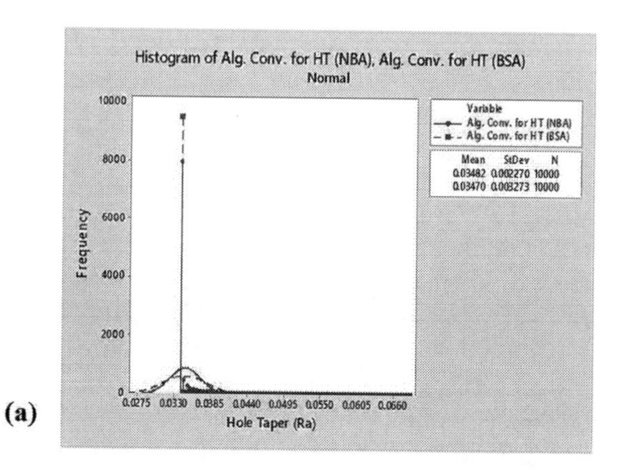

(b)

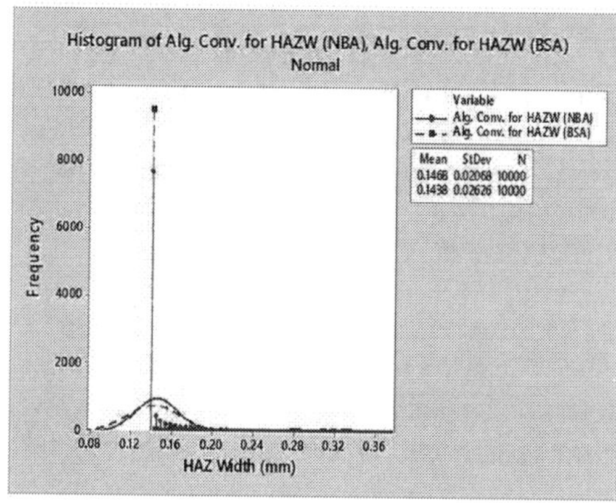

FIGURE 5.1
Comparisons of functional evaluations for NBA and BSA with respect to hole taper and HAZ width by means of histogram of functional evaluations

TABLE 5.4

Performance evaluation of NBA and BSA for single objective optimization

Algorithm	Functional evaluations	Objective function	Convergence speed	No. of optimized population	Computational time (in seconds)	
NBA	10,000	HT	Converged after 185 evaluations	7784	0.3907	0.394
		HAZW	Converged after 549 evaluations	6724	0.3964	
BSA	10,000	HT	Converged after 87 evaluations	9470	0.3130	0.318
		HAZW	Converged after 63 evaluations	9485	0.3239	

computational times observed for single-response optimization using NBA and BSA are 0.394 and 0.318 seconds, respectively.

5.4.2 Multi-objective Optimization

In this segment, both hole taper and HAZ width are optimized instantaneously using NBA and BSA. The following objective function is developed for carrying out multi-objective optimization:

$$minZ = \frac{w_1 \times HT}{HT_{min}} - \frac{w_2 \times HAZW}{HAZW_{min}} \tag{5.21}$$

where w_1 and w_2 are the allocated weights to hole taper and HAZ width, correspondingly (so as to $w_1 + w_2 = 1$). The min values in the denominator are taken from the results of single response optimization of laser micro-drilling using NBA and BSA. Table 5.5 shows the results of multi-response optimization based on the selected criteria. It is seen from Table 5.5 that both NBA and BSA outperform the optimal value predicted by RSM, and gives effective optimal parametric setting.

Figure 5.2 shows the histograms of algorithm convergence data for NBA and BSA with respect to multi objective function (Z_{min}). It is evident from Figure 5.2 that BSA shows faster convergence than NBA. It is seen from Figure 5.2 that the mean value of the results obtained by BSA (1.032 for Z_{min}) are closer to the optimum results (0.9998 for Z_{min}), compared to the mean values obtained by NBA (1.036 for Z_{min}), which signifies that more number of evaluations have converged to its optima for BSA as compared to NBA. Table 5.6 provides the algorithm convergence report details in terms of convergence speed and number of optimized populations for multi-objective optimization. The NBA converges to the optima after 367 evaluations, whereas only 209 evaluations have been taken by BSA. For NBA, almost 7,338 evaluations have been found converged to its optimal value, whereas for BSA it is 9,319 evaluations. The average computational times observed for multi response optimization using NBA and BSA are 0.413 and 0.333 seconds, respectively.

TABLE 5.5

Results of multi-objective optimization

Conditions	Optimization technique	Z_{min}	Optimal value		Optimal parameter setting			
			Hole taper	HAZ width	*A*	*B*	*C*	*D*
$w_1 = w_2 = 0.5$	NBA	0.9998	0.0341	0.1389	20.00	5.00	0.60	18.00
	BSA	0.9998	0.0341	0.1389	20.00	5.00	0.60	18.00
	RSM	–	0.0460	0.2596	21.00	3.93	1.01	14.00

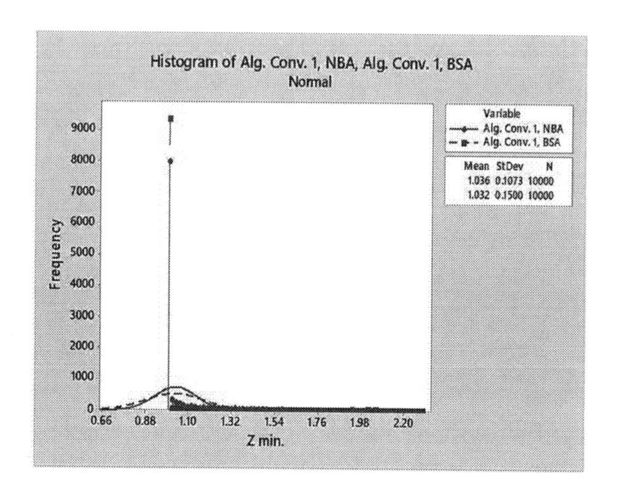

FIGURE 5.2
Comparisons of functional evaluations for NBA and BSA with respect to multi-objective function (Z_{min})

TABLE 5.6
Performance evaluation of NBA and BSA for multi-objective optimization

Algorithm	Functional evaluations	Objective function	Convergence speed	No. of optimized population	Computational time (in seconds)
NBA	10,000	Z_{min}	Converged after 367 evaluations	7,338	0.413
BSA	10,000	Z_{min}	Converged after 209 evaluations	9,319	0.333

5.5 Prediction of Parametric Trends Using NBA and BSA

Figure 5.3 furnishes the scatter plot to show the variations of hole taper with process parameters. NBA and BSA can both predict true overall parametric trends, as no parameters are required to kept constant during analysis, and thus it is advantageous than RSM. It has been clearly observed from the actual scatter plot and linearly fitted plot of Figure 5.3(a) and (c) that hole taper increases with lamp current and air pressure. High thermal energy is generated at higher lamp current, resulting in a large taper. It is clear from Figure 5.3(b) and (d), with an increase in pulse frequency and pulse width, the hole taper decreases. High beam energy is generated at low pulse frequencies, which produce tapered holes. Figure 5.4 shows the trends of HAZ width with respect to the process parameters. It is seen from Figure 5.4 that increasing lamp current and air pressure, increases HAZ

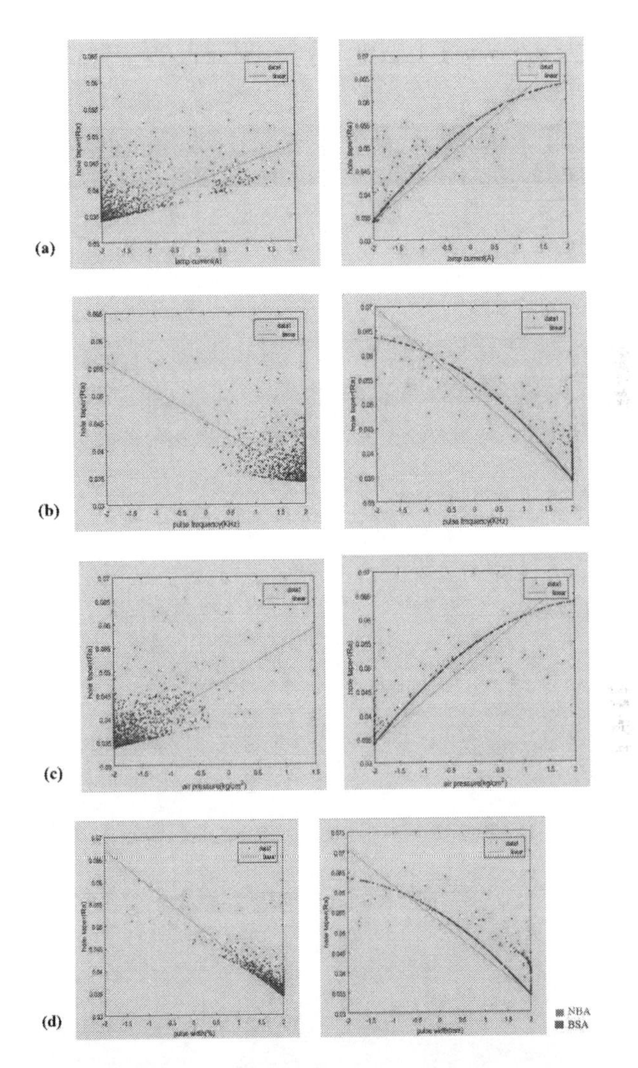

FIGURE 5.3
Parametric trends of hole taper for (a) lamp current, (b) pulse frequency, (c) air pressure and (d) pulse width

width, whereas, an increase of pulse frequency and pulse width decreases HAZ width. The high lamp current produces high HAZ width because of the generation of high thermal energy at higher values of lamp currents. At higher level of assist air pressures, low HAZ width is observed, because higher air pressures aid to quick ejection of the molten material and the removal of excess heat. It is evident from Figure 5.3 and Figure 5.4 that BSA maps a sharp trend line while NBA gives an approximate trend with scattered data points.

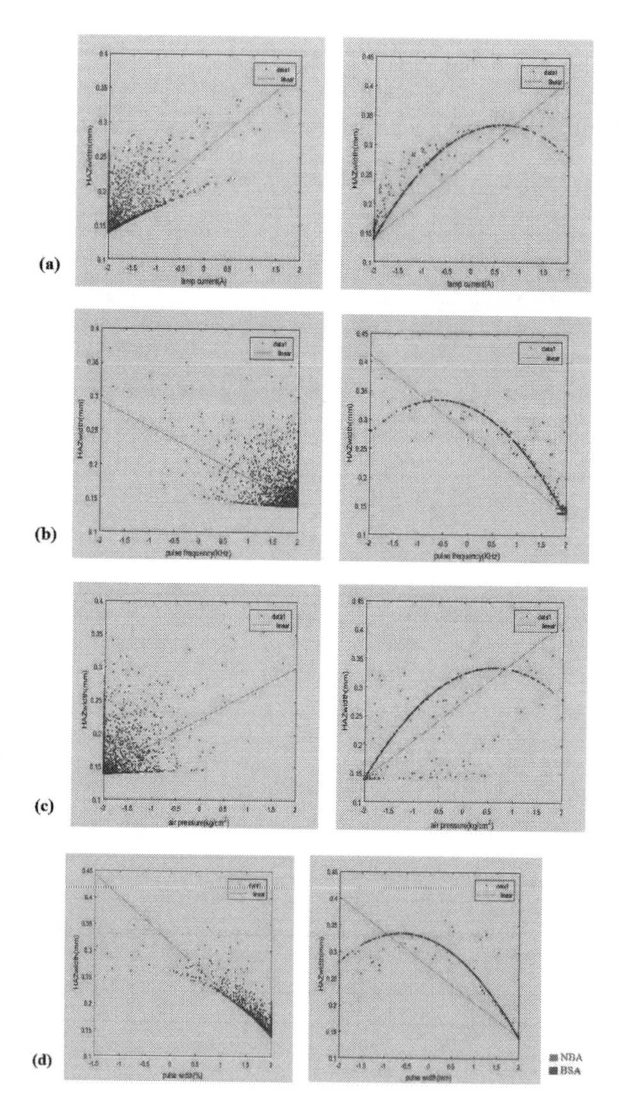

FIGURE 5.4
Parametric trends of HAZ width for (a) lamp current, (b) pulse frequency, (c) air pressure and (d) pulse width

5.6 Conclusion

Noble bat algorithm and bird swarm algorithm are implemented for optimization of laser micro-drilling parameters and for prediction of parametric trends. The micro-drilled hole qualities, namely hole taper and HAZ width, are optimized with respect to input parameters like lamp current, air pressure, pulse frequency and pulse width. The results obtained using NBA and BSA is compared with the results of RSM, which demonstrates the effectiveness of NBA and BSA for optimization of laser micro-drilling parameters to enhance the quality characteristics. It is observed that BSA provides more effective and better result than NBA in terms of convergence speed and number of optimized populations. The average computational times observed for NBA and BSA are less than a second. BSA takes less time than NBA to solve single as well as multi-objective optimization problems. It is further seen that NBA and BSA can predict the true overall parametric trends because they do not require keeping any process parameter as constant during the analysis.

References

Biswas, R., Kuar, A. S., & Mitra, S. (2015). Process optimization of Nd: YAG laser micro drilling of alumina-aluminium interpenetrating phase composite. *Journal of Materials Research and Technology, 4*(3), 323–332.

Du, Z. Y., & Liu, B. (2012). Image matching using a bat algorithm with mutation. *Applied Mechanics and Materials, 203*, 88–93.

Gautham, S., & Rajamohan, J. (2016). Economic load dispatch using novel bat algorithm. *Proceeding of IEEE International Conference on Power Electronics, Intelligent Control and Energy Systems (ICPEICES)*, July 4–6, Delhi, India.

Ghoreishi, M., Low, D.K.Y., & Li, L. (2002). Comparative statistical analysis of hole taper and circularity in laser percussion drilling. *International Journal of Machine Tools and Manufacture, 42*, 985–995.

Goel, H., & Pandey, P. M. (2014). Experimental investigations into micro-drilling using air assisted jet electrochemical machining. *Proceeding of AIMTDR*, December 12th–14, IIT Guwahati, Assam, India.

Jackson, M. J., & O'Neill, W. (2003). Laser micro-drilling of tool steel using Nd: YAG lasers. *Journal of Material Processing Technology, 142*, 517–525.

Khuri, A. I., & Cornell, J. A. (1996). *Response surface: Design and analysis.* New York: Marcel Dekker.

Kuar, A. S., Acherjee, B., & Mitra, S. (2012). Laser micro-drilling of alumina: Parametric modeling and sensitivity analysis. *International Journal of Mechatronics and Manufacturing Systems, 5*(3/4), 294–307.

Kuar, A. S., Paul, G., & Mitra, S. (2006). Nd: YAG laser micro machining of alumina—aluminium interpenetrating phase composite using response surface methodology. *International Journal of Machining and Machinability of Materials, 1*(4), 423–444.

Liu, W. (2016). The application research of improved bat algorithm for time table problem. *Proceeding of 4th International Conference on Machinery, Materials and Computing Technology (ICMMCT 2016)*, Atlantis Press, pp. 1667–1670.

Meng, X. B., Gao, X. Z., Liu, Y., & Zhang, H. (2015). A novel bat algorithm with habitat selection and Doppler effect in echoes for optimization. *Expert Systems with Applications, 42*, 6350–6364.

Meng, X.B., Gao, X.Z., Lu, L., Liu, Y., & Zhang, H. (2016). A new bio-inspired optimisation algorithm: Bird swarm algorithm. *Journal of Experimental & Theoretical Artificial Intelligence, 28*, 673–687.

Montgomery, D.C. (2001). *Design and analysis of experiments*, 5th Edition. New York: John Wiley & Sons, Inc.

Musikapun, P., & Pongcharoen, P. (2012). Solving multi-stage multi-machine multi-product scheduling problem using bat algorithm. *Proceeding of 2nd International Conference on Management and Artificial Intelligence (IPEDR)*, IACSIT Press, Singapore, *35*, pp. 98–102.

Parashar, M., Rajput, S., & Dubey, H.M. (2017). Optimization of benchmark functions using a nature inspired bird swarm algorithm. *Proceeding of 2017 3rd International Conference on Computational Intelligence & Communication Technology (CICT)*, February 9–10, Ghaziabad, India.

Sambariya, D.K., & Prasad, R. (2014). Robust tuning of power system stabilizer for small signal stability enhancement using metaheuristic bat algorithm. *International Journal of Electrical Power & Energy Systems, 61*, 229–238.

Yang, X.S. (2010). A new metaheuristic bat-inspired algorithm, in Nature inspired cooperative strategies for optimization (NICSO 2010). *Studies in Computational Intelligence*. Springer Verlag, Berlin, *284*, pp. 65–74.

Yang, X.S. (2013). Bat algorithm: Literature review and applications. *International Journal of Bio-Inspired Computation, 5*(3), 141–149.

Yang, X.S. (2013). Bat algorithm: Literature review and applications. *International Journal of Bio-Inspired Computation, 5*(3), 141–149.

Yilbas, B.S., & Yilbas, Z. (1987). Parameters affecting hole geometry in laser drilling of Nimonic 75. *Proceedings of SPIE, 744*, 87–91.

Section III

Application to Energy Systems

6

Energy Demand Management of a Residential Community through Velocity-Based Artificial Colony Bee Algorithm

Sweta Singh

Research Scholar, Dept. of Electrical Engineering, Manipal University Jaipur, Jaipur, India, Email: swetasingh1231@gmail.com

Neeraj Kanwar

Assistant Professor, Dept. of Electrical Engineering, Manipal University Jaipur, Jaipur, India, Email: nk12.mnit@gmail.com

CONTENTS

6.1 Introduction 103
6.2 Different Loads for Residential Community 106
6.3 Problem Formulation 109
6.4 Implementation of Velocity-Based Artificial Colony Bee Algorithm 109
6.5 Results and Discussion 110
6.6 Conclusion 114
References 115

6.1 Introduction

In today's scenario of rapid economic development, there has been an exponential rise in the demand of energy both from residential as well as commercial consumers. As such, energy management and hence its effective utilization has gained centre stage within the research domain. Portion of the load is now being adjusted through the utilization of sustainable energy sources. The overall objective is to reduce CO_2 emissions and hence reduce the possible impact on global temperature. However, there have been increased uncertainties in the supply of power from the grid owing to the inclusion of renewable energy sources. Furthermore, the conventional power supply scheme has changed tremendously. There has been additional stress on the grid apart from that presented by the power system deregulations. Therefore, the control and operation of power system has become more challenging than ever it was. To take care of the stress aspects, the smart grid is one of the established mechanisms that provides the grid with enhanced reliability, security, capacity and efficiency (Hernández et al. 2012). The major driving force behind transforming the grid to smarter ones is the reformed structure of electricity as well as the associated challenges. The scope of smart grid operation encompasses large-scale operation of power system components as well as their control to the distribution networks that handles the bulk generation from the power utility. However, the reliability of the system operation depends to a greater extent on the real-time information sharing between the central power generators as well as the end-consumers of electricity. The effective communication system aids in taking reliable, secure and economic decisions dynamically. Gungor et al. (2011) have discussed the importance of different communication technologies as well as the associated challenges. These challenges are critical to be addressed so that an effective grid environment can be realized.

Participation of the end-consumer is also another dimension of the smart grid network. This aids in improving the demand response which can also be underpinned through demand side management (DSM) architecture. Since DSM architecture benefits both the consumers as well as the utility, it is much more effective than the supply side management techniques. The various reasons to the development of DSM architecture are the problems associated with peak load, distribution deficiencies, transmission deficiencies, environmental concerns etc. the load profile is improved through the effective design and planning of the end-consumer activities. Tersely, one can deduce DSM to be an effective mechanism that controls the consumers' electricity consumption. There have been studies regarding this aspect in the research domain. Incentive-based autonomous schedules of energy consumption were addressed in Mohsenian-Rad et al. (2010). The formulation was made to minimize the cost of electricity usage. In another study, an industrial project for supplying in-line hot water to the consumers was discussed (Rankin and Rousseau 2008). Through the study it was revealed that the DSM architecture not only provides strategies to reschedule the consumer electricity demand but also facilitate the optimal operation of the installed renewable energy generators. The optimal operation of renewable energy sources was described in (Finn et al. 2013) and revealed the financial savings of the end-consumers.

DSM architecture has been effective in controlling the thermostatically customer loads which includes air conditioners and refrigerators. The inside cooling is accomplished when the compressor of the associated load is switched on, and when the desired temperature has been achieved, the compressor is switched off. The DSM application can aid in controlling the cooling activities either by pre-cooling or delayed cooling. The economic aspects associated with cooling loads have been depicted in the study carried out by (Zehir and Bagriyanik 2012). However, accurate mathematical modelling of the cooling loads is quintessential to achieve an effective scheduling of the thermostatically loads (Perfumo et al. 2012).

The energy management program from the energy utilities are aimed to improve the operational efficiency. This is accomplished by encouraging the consumers to adopt load schedule that produces optimal results in terms of power consumption. There are certain factors that should be kept in mind while doing so: customer requirement, their comfort level and the price of the energy distributed. Household load scheduling has been accomplished in one of the studies (Du and Lu 2011), in which case the algorithm requires energy price as well as the scenarios of energy consumption forecast. The objective may be either to minimize the electricity bill or to maximize the comfort level of the consumer. Through the adoption of optimal operation and management policies, the building owners as well as the residential customers can save quantum of their electricity bills without affecting significantly the involved infrastructure. Traditional buildings can be transformed to smart a building that adopt the strategies associated with optimal scheduling and also coordinates the activities associated with the operation of renewable energy sources (Guan et al. 2010). A discussion on energy storage and their impacts on the electricity supply pattern have been carried out in Qureshi et al. (2011). Load management strategies have been designed to control the demands of different consumers buying energy from power utility. Different control and modifications techniques enable the power utility to meet the power requirements of their consumers in an economic way (Kostková et al. 2013). Different tariff schemes have been deployed by several countries to reduce the consumer electricity bill, such as time of use (TOU) and day-ahead-price tariffs (Godoy-Alcantar and Cruz-Maya 2011). DSM strategy for commercial customers under a TOU tariff scheme was proposed by Malakar et al. (2019).

In the present work, a DSM strategy has been proposed to minimize the total cost of electricity for a residential society. The residential society under consideration has both controllable as well as uncontrollable loads. The formulated cost objective function has been minimized using the velocity-based artificial colony bee algorithm (VABC). A number of constraints have been inflicted by both the controllable as well as uncontrollable loads. The effectiveness of the DSM strategy has been adjudged by obtaining per unit cost of electricity consumption. A comprehensive load management and optimization approach has been evidenced through the simulation results. The total electricity bill was found to be reduced with the effective DSM strategy in place.

6.2 Different Loads for Residential Community

Owing to the computational constraints, it is not feasible to involve smaller loads directly to participate in the wholesale energy markets. An aggregator is usually employed that serves to interface the small consumers of electricity with the existing market. A typical aggregation scheme comprises a number of customers and an aggregator. The demand within the aggregation scheme is controlled via automated energy management systems. The aggregator in place formulates an effective control policy that can optimally control and schedule the part loads of the participating users. To accomplish the task, aggregator communicates with the automated energy management system directly. The aggregator provides signals which represent the energy price per unit interval or by sending the control signals to the different devices. In the present study, the framework comprises a number of residential customers, the power utility and the solar photovoltaic energy generators.

There are certain numbers of controllable as well as uncontrollable loads associated with every residential consumer (Nan et al. 2018). Uncontrollable appliances without storage are the ones that don't store energy and are required to function until and unless the desired work is completed. Some of the prominent examples are personal computers, televisions, lamps and so forth. Controllable appliances without storage are the appliances that don't store energy, however their consumption of energy can be controlled. Washing machines, dryers and dishwashers are some of the instances of such appliances. The appliances fall under the category of controllable appliances with direct storage if they have the potential ability to store energy and their energy consumption can be controlled flexibly. Plug-in hybrid vehicles are examples of such appliances. There are controllable appliances with indirect storage capabilities. Such appliances transfer the electric energy to other forms of energy. Examples include water heaters, air conditioning systems and so forth.

The different loads used by residential customer have been tabulated in Table 6.1.

Apart from the small loads of the appliances, there are bigger loads in a residential community such as lighting load, basic load, lift load and pumping load. The lighting loads as well as the lift loads have been depicted in Figure 6.1 and Figure 6.2; the lift load depends on the number of persons in the lift. The energy is required to run the alarm system, fire systems and so forth in the residential community. The basic load has been estimated to be 25 kW.

Furthermore, the society has pumping system to load water from the underground storage facility to the water storage tank at the top of the building. The pumping load has been approximated to be 145 kW.

Therefore, the total load can be derived using equation 6.1 below:

$$L_l(t) = \sum_{R=1}^{n}\sum_{k=1}^{c} x_{R,c}^{k}(t) + \sum_{R=1}^{n}\sum_{m=1}^{uc} x_{R,uc}^{m}(t) \tag{6.1}$$

TABLE 6.1
Residential load data

Load type	Power rating
Rice cooker	0.2
Microwave oven	1.5
Coffee maker	0.5
Electric iron	1
Personal computers	0.2
Television	0.2
Ceiling fan	0.07
Dishwasher	1.2
Electric kettle	0.75
Electric shaver	0.02
Fridge	0.4
Gaming PC	0.5
Home air conditioner	3
Induction cooktop	2
Washing machine	0.5

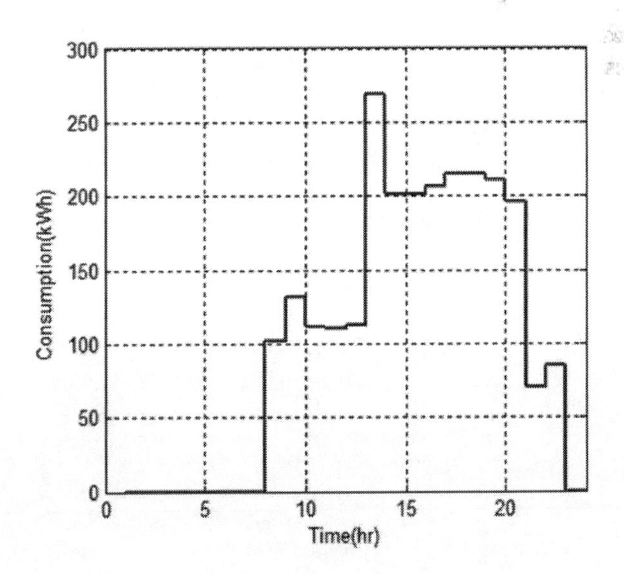

FIGURE 6.1
Energy consumption by lift

where L is the total load of the system, $x_{R,c}^{k}(t)$ is the consumption of kth controllable appliance of Rth customer at any time interval t and $x_{R,uc}^{m}(t)$ is the consumption of mth controllable appliance of Rth customer at any time interval t.

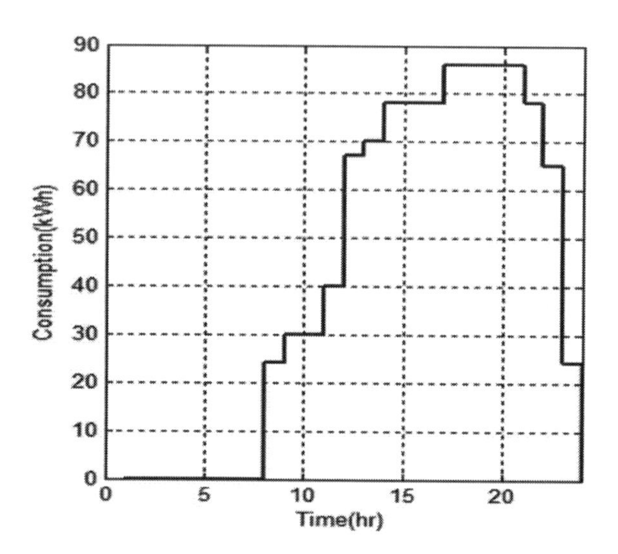

FIGURE 6.2
Energy consumption by lighting system

The residential community under consideration also has a solar photovoltaic generation system that transforms the solar energy to electric energy. The power output of a solar PV system depends on the solar irradiation, atmospheric temperature, the efficiency of the convertor and the PV array. It has been assumed that a maximum power point tracker has been installed with the solar PV system to extract maximum possible power. The output power of a solar PV (Akram and Khalid 2018) can be calculated using equation 6.2.

$$P_{pv}(t) = \eta_{pv} A_{pv} I(t)(1 - 0.005(T_o(t) - 25)) \tag{6.2}$$

where L_{pv} is power output of the solar system in kW, area of PV array is denoted by A_{pv} in m^2, T_o is the atmospheric temperature in °C, I is the solar irradiation in kW/m^2. The efficiency of the solar PV system is denoted by η_{pv}.

Equation 6.3 can be used to obtain the load supplied by the solar generation system:

$$L_{pv}(t) = L_l(t) \times 1_{\{P_{pv}(t) \geq L_l(t)\}} + P_{pv}(t) \times 1_{\{P_{pv}(t) < L_l(t)\}} \tag{6.3}$$

It is also compressible that the residential customer purchases electricity from the power utility only when their load demand is not met by the installed solar photovoltaic system. The energy supplied by the power utility can be determined using equation 6.4:

$$P_u(t) = \{L_l(t) - P_{pv}(t)\} \times 1_{\{P_{pv}(t) < L_l(t)\}} \tag{6.4}$$

6.3 Problem Formulation

The main objective of the present work is to minimize the total electricity cost by shifting the loads to less pricing hours and by utilization of the solar PV system whenever available. The total cost of energy supplied by the power utility can be expressed as $\sum P_u(t) \times C(t)$. The discounted energy served is denoted as $\sum L_{pv}(t) \times [\{(1+d)^Y - 1\} / \{d(1+d)^Y]$.

The objective function can be formulated and is expressed in equation 6.5:

$$Minimize \sum_{t=1}^{T} \sum_{L=1}^{l} (L_t^l C_t) \tag{6.5}$$

Subject to the following constraints:

$$L_t - L_{pv} - P_u = 0 \tag{6.6}$$

$$\sum_{L=1}^{l} (Y_l^t L_l^t) \leq E_{\max} \tag{6.7}$$

$$L_l^t \leq L_l^{\max} \tag{6.8}$$

Equation 6.6 signifies the power balance equation. The total energy consumed by any load must be limited during the entire day operation. The restriction on total load is depicted in equation 6.7. Y_l^t denotes the load activity and the quantity for any load is represented by L_l^t. The maximum capacity at which any load must be operated is represented by E_{max}. Equation 6.8 represents the mathematically the maximum operation aspect.

In the present work it has also been assumed that the solar PV sells power to the utility grid in case the load generated is more than that required. This results in maximizing the total profit. The total cost of energy sold to the utility is subtracted from the total cost.

6.4 Implementation of Velocity-Based Artificial Colony Bee Algorithm

The artificial colony bee (ABC) algorithm is one of the metaheuristics that falls under the diversified categories of swarm intelligence techniques and was developed by Karaboga (2005). The ABC algorithm stimulates the behaviour of honeybees. When compared to conventional techniques such as PSO and GA, ABC has number of associated advantages. There is a higher probability of finding the optimal solution, as the ABC algorithm implements a global as well as a local search at each iterations. It employs three phases to search for the optimal solution: employed bee stages, onlooker bee phase and finally the scouting bee phase. The ABC algorithm has been employed in variety of optimization problems of various types: constrained and non-constrained (Chang 2013; Karaboga and Latifoglu 2013; Das 2013; Tsai 2014; Zhang 2014; Szeto and Jiang 2014; Li et al. 2014).

However, one of the challenging aspects associated with the ABC algorithm is its slower convergence speed in comparison to other algorithms such as differential evolution and PSO. Furthermore, the algorithm gets trapped in local optima while solving more complex problems (Karaboga and Akay 2009). Hence ABC has poor exploitation ability while its exploration capability is effective. Therefore in order to suppress the aforementioned drawbacks, few modified version of ABC algorithm have been proposed. These include best so far ABC (Banharnsakun et al. 2013), ERABC (Xiang and An 2013), HPA (Kiran and Gündüz 2013), SAABC (Chen et al. 2012), COABC (Gao and Liu 2012, 2019) and so on.

The present work solves the objective function through the velocity-based artificial bee colony (VABC) optimization algorithm (Imanian et al. 2014). VABC is a hybridized version of traditional ABC algorithm with PSO. The VABC also finds the optimal solution in three stages: the first phase employs the exploration property of ABC to search the space defined by the problem under consideration. The second phase employs PSO to find optimal solution of each particle. In the final phase of the VABC algorithm, the best solutions are maintained whereas the other solutions are replaced with some random solutions.

The VABC algorithm has been implemented through MATLAB. The results obtained with the algorithm implementation have been found to be effective.

6.5 Results and Discussion

The simulated total load is shown in Figure 6.3. The load demand is embraced of controllable and uncontrollable appliances. On the basis of energy consumption in the residential sector, the daily peak load appears between 7 and 9 p.m. because of the fact that most of the customers reach home and fulfil their needs. Per unit cost electricity is shown in Figure 6.4. According to the figure the maximum cost of energy appears between 8 and 10 p.m. and minimum cost between 2 and 5 p.m. It can be analyzed from Figure 6.2 and Figure 6.3 that during maximum demand of energy per unit cost is high and during less demand the per unit cost is low. The research work

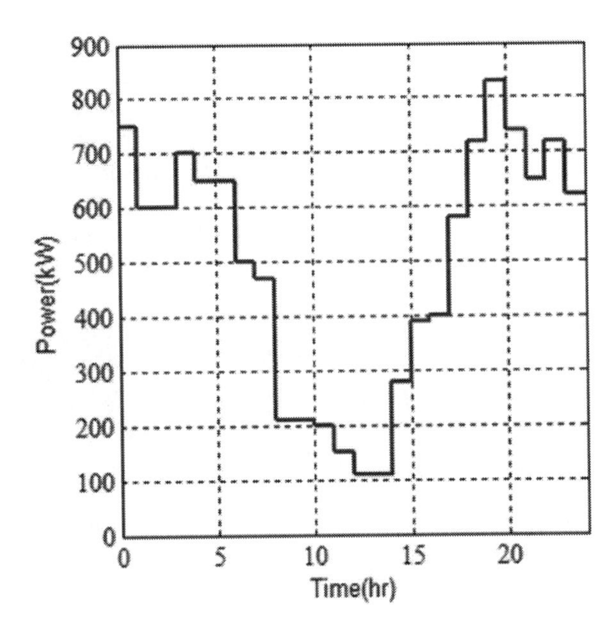

FIGURE 6.3
Simulated total load

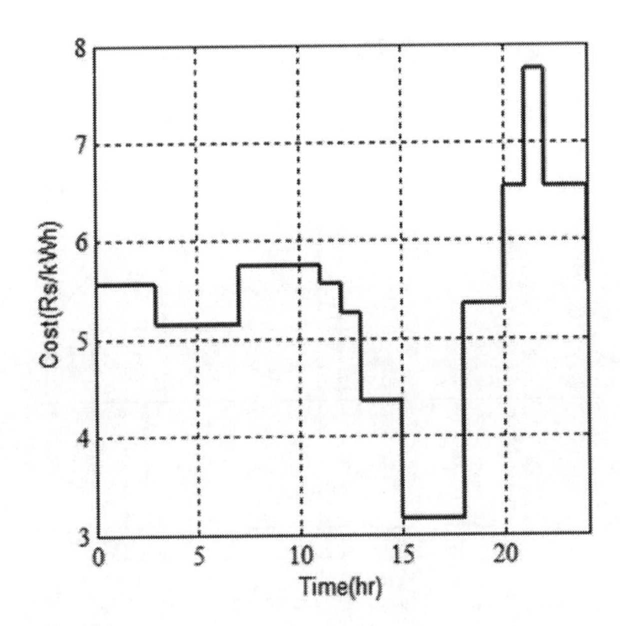

FIGURE 6.4
Per unit cost of electricity

has been done initially without DSM for normal load then with DSM which optimizes the load and further DSM program along with self-generation.

The simulation has been done on the system for normal load without DSM and the value of cost per unit is calculated approximately Rs. 5.454 per kWh. For the same system in the presence of demand side management that optimize the load with day ahead-pricing by the utility, per unit cost is calculated which is around Rs. 5.017 per kWh. The price reduction is due to demand side management program which shift the load in off-peak hour or low pricing time for gaining the system constraints.

Further analysis has been done with the purpose to achieve minimum cost of energy PV generation system is installed. The cost function developed using VABC and best result has been found with the help of algorithm. It minimizes the cost function with affecting the comfort of users.

The optimal size of photovoltaic generation is 250 kW. The output of PV generation is shown in Figure 6.5. It is analyzed that the PV generation is maximum in day-ahead timing and it is maximum between 12 and 2 p.m. and then decreasing eventually. The system is operated with DSM receive energy from PV generation and per unit cost of electricity become Rs. 4.615 per kWh. The system operation in different condition for a day is shown in Figure 6.6. If the local generation is less to fulfil the demand, it buys from the utility. Utility energy given to the users is shown in Figure 6.7. In condition of surplus energy generation by the local generation sells excessive power to the utility, which is shown in Figure 6.8.

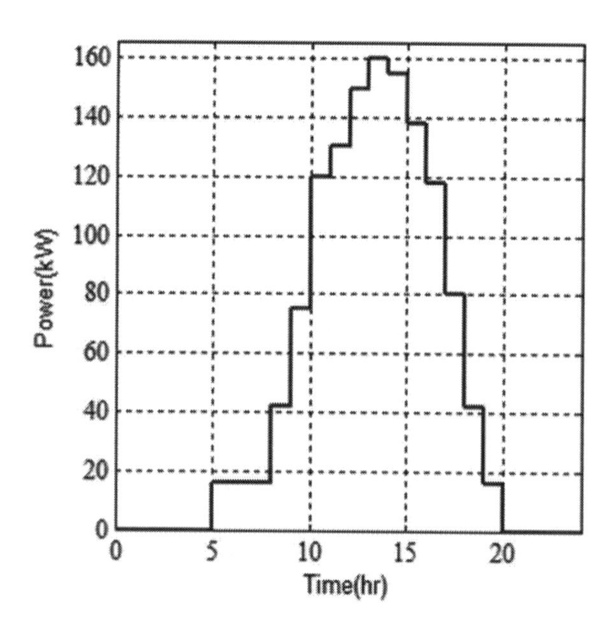

FIGURE 6.5
Photovoltaic system served to the load

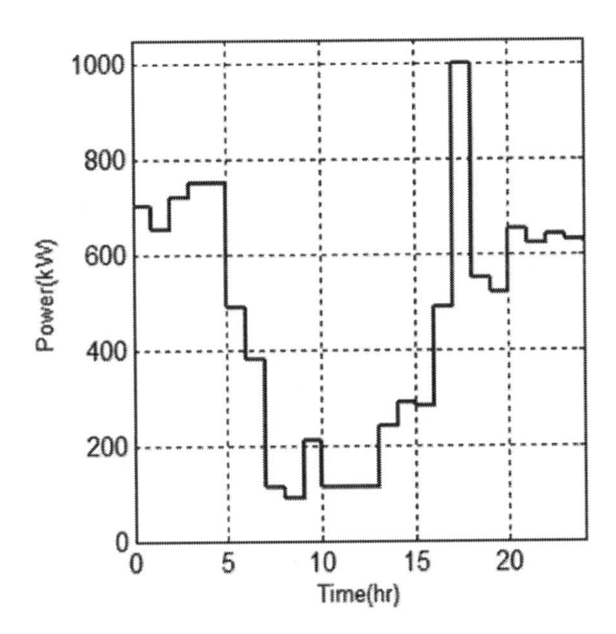

FIGURE 6.6
Shifted load consumption

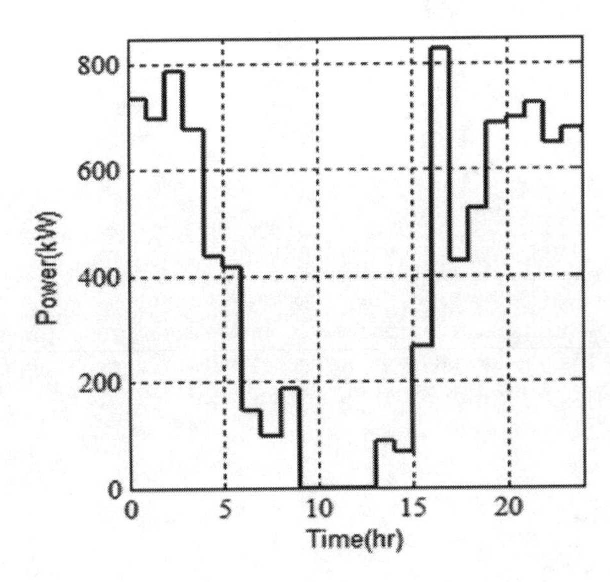

FIGURE 6.7
Load provided by grid system

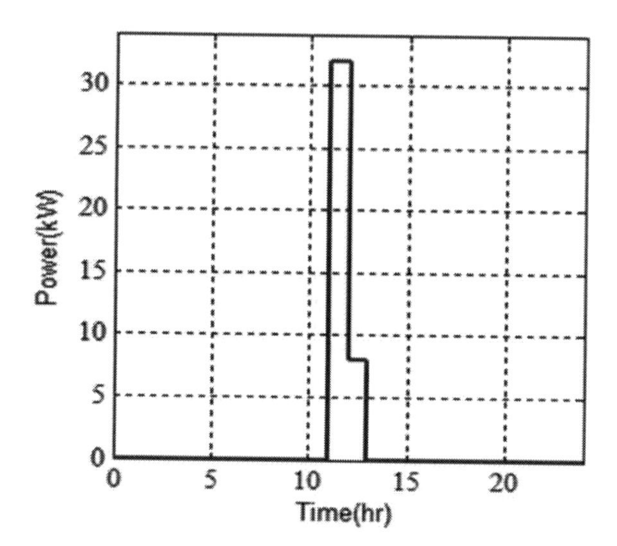

FIGURE 6.8
Energy sold by consumer to grid system

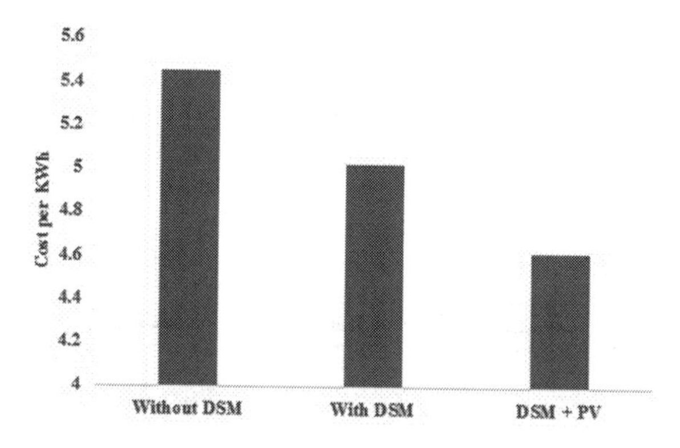

FIGURE 6.9
Simulated per unit cost of energy in three conditions

Based on per unit cost a comparison is shown in Figure 6.9 for energy consumption by the residential users. It can be easily analyzed that per unit cost of energy is at a minimum when PV is integrated with the DSM program.

6.6 Conclusion

In the present work, DSM strategy has been proposed to minimize the electricity cost of a residential society. The proposed DSM strategy optimizes the pattern of electricity consumption and hence brings down the total cost of electricity. The velocity-based artificial colony bee algorithm was employed to implement the DSM strategy, and it was revealed that the cost of electricity was least when the DSM strategy was in place together with the solar PV system. Electric vehicle charging loads could be added in further studies as a future work to the present proposed work.

References

Akram, U., & Khalid, M. (2018). Residential demand side management in smart grid paradigm. *2018 IEEE PES Asia-Pacific Power and Energy Engineering Conference (APPEEC)*, October, IEEE, pp. 439–444.

Banharnsakun, A., Sirinaovakul, B., & Achalakul, T. (2013). The best-so-far ABC with multiple patrilines for clustering problems. *Neurocomputing, 116*, 355–366.

Chang, W.D. (2013). Nonlinear CSTR control system design using an artificial bee colony algorithm. *Simulation Modelling Practice and Theory, 31*, 1–9.

Chen, S.M., Sarosh, A., & Dong, Y.F. (2012). Simulated annealing based artificial bee colony algorithm for global numerical optimization. *Applied Mathematics and Computation, 219*(8), 3575–3589.

Das, S., Biswas, S., & Kundu, S. (2013). Synergizing fitness learning with proximity-based food source selection in artificial bee colony algorithm for numerical optimization. *Applied Soft Computing, 13*(12), 4676–4694.

Du, P., & Lu, N. (2011). Appliance commitment for household load scheduling. *IEEE Transactions on Smart Grid, 2*(2), 411–419.

Finn, P., O'Connell, M., & Fitzpatrick, C. (2013). Demand side management of a domestic dishwasher: Wind energy gains, financial savings and peak-time load reduction. *Applied Energy, 101*, 678–685.

Gao, W., Liu, S., & Huang, L. (2012). A global best artificial bee colony algorithm for global optimization. *Journal of Computational and Applied Mathematics, 236*, 2741–2753.

Godoy-Alcantar, J.M., & Cruz-Maya, J.A. (2011). Optimal scheduling and self-generation for load management in the Mexican power sector. *Electric Power Systems Research, 81*(7), 1357–1362.

Guan, X., Xu, Z., & Jia, Q.S. (2010). Energy-efficient buildings facilitated by microgrid. *IEEE Transactions on Smart Grid, 1*(3), 243–252.

Gungor, V.C., Sahin, D., Kocak, T., Ergut, S., Buccella, C., Cecati, C., & Hancke, G.P. (2011). Smart grid technologies: Communication technologies and standards. *IEEE Transactions on Industrial Informatics, 7*(4), 529–539.

Hernández, L., Baladrón, C., Aguiar, J., Carro, B., & Sánchez-Esguevillas, A. (2012). Classification and clustering of electricity demand patterns in industrial parks. *Energies, 5*(12), 5215–5228.

Imanian, N., Shiri, M.E., & Moradi, P. (2014). Velocity based artificial bee colony algorithm for high dimensional continuous optimization problems. *Engineering Applications of Artificial Intelligence, 36*, 148–163.

Karaboga, D. (2005). An idea based on honey bee swarm for numerical optimization, Technical Report.

Karaboga, D., & Akay, B. (2009). A comparative study of artificial bee colony algorithm. *Applied Mathematics and Computation, 214*(1), 108–132.

Karaboga, N., & Latifoglu, F. (2013). Elimination of noise on transcranial Doppler signal using IIR filters designed with artificial bee colony—ABC-algorithm. *Digital Signal Processing, 23*(3), 1051–1058.

Kıran, M.S., & Gündüz, M. (2013). A recombination-based hybridization of particle swarm optimization and artificial bee colony algorithm for continuous optimization problems. *Applied Soft Computing, 13*(4), 2188–2203.

Kostková, K., Omelina, L., Kyčina, P., & Jamrich, P. (2013). An introduction to load management. *Electric Power Systems Research, 95*, 184–191.

Li, J.Q., Pan, Q.K., & Tasgetiren, M.F. (2014). A discrete artificial bee colony algorithm for the multi-objective flexible job-shop scheduling problem with maintenance activities. *Applied Mathematical Modelling, 38*(3), 1111–1132.

Malakar, T., Goswami, S.K., & Rajan, A. (2019). Demand side management of a commercial customer based on ABC algorithm. *Soft Computing for Problem Solving*, Springer, Singapore, pp. 617–634.

Mohsenian-Rad, A.H., Wong, V.W., Jatskevich, J., Schober, R., & Leon-Garcia, A. (2010). Autonomous demand-side management based on game-theoretic energy consumption scheduling for the future smart grid. *IEEE Transactions on Smart Grid, 1*(3), 320–331.

Nan, S., Zhou, M., & Li, G. (2018). Optimal residential community demand response scheduling in smart grid. *Applied Energy, 210*, 1280–1289.

Perfumo, C., Kofman, E., Braslavsky, J.H., & Ward, J.K. (2012). Load management: Model-based control of aggregate power for populations of thermostatically controlled loads. *Energy Conversion and Management, 55*, 36–48.

Qureshi, W.A., Nair, N.K.C., & Farid, M.M. (2011). Impact of energy storage in buildings on electricity demand side management. *Energy Conversion and Management, 52*(5), 2110–2120.

Rankin, R., & Rousseau, P.G. (2008). Demand side management in South Africa at industrial residence water heating systems using in line water heating methodology. *Energy Conversion and Management, 49*(1), 62–74.

Szeto, W.Y., & Jiang, Y. (2014). Transit route and frequency design: Bi-level modeling and hybrid artificial bee colony algorithm approach. *Transportation Research Part B: Methodological, 67*, 235–263.

Tsai, H.C. (2014). Integrating the artificial bee colony and bees algorithm to face constrained optimization problems. *Information Sciences, 258*, 80–93.

Xiang, W.L., & An, M.Q. (2013). An efficient and robust artificial bee colony algorithm for numerical optimization. *Computers & Operations Research, 40*(5), 1256–1265.

Zehir, M.A., & Bagriyanik, M. (2012). Demand side management by controlling refrigerators and its effects on consumers. *Energy Conversion and Management, 64*, 238–244.

Zhang, D.L., Ying-Gan, T.A.N.G., & Xin-Ping, G.U.A.N. (2014). Optimum design of fractional order PID controller for an AVR system using an improved artificial bee colony algorithm. *Acta Automatica Sinica, 40*(5), 973–979.

7

Adaptive Neuro-fuzzy Inference System (ANFIS) Modelling in Energy System and Water Resources

P. A. Adedeji

Department of Mechanical Engineering Science, University of Johannesburg, South Africa

S. O. Masebinu

Department of Mechanical Engineering Science, University of Johannesburg, South Africa

S. A. Akinlabi

Department of Mechanical and Industrial Engineering, University of Johannesburg, South Africa

Department of Mechanical Engineering, Covenant University, Ota, Nigeria

N. Madushele

Department of Mechanical Engineering Science, University of Johannesburg, South Africa

CONTENTS

7.1 Introduction 118
7.2 Time Series Data in Energy and Water Resources 119
7.3 Black Box Models 120
 7.3.1 Performance Evaluation of Black Box Models 121
7.4 Fuzzy Inference System 124
7.5 Adaptive Neuro-fuzzy Inference System (ANFIS) 124
7.6 Peculiarities of ANFIS Model 126
7.7 Case Studies in Energy and Water Resource Modelling 127
 7.7.1 ANFIS Models in Energy and Water System 128
 7.7.2 Hybrid ANFIS in Energy and Water Resource Modelling 129
7.8 Conclusion 130
References 131

7.1 Introduction

Unravelling the complexity of the energy, water, and food nexus has lately become a growing concern among researchers, practitioners and policy makers (Smajgl et al. 2016; Cai et al. 2018). Among these three, there exist two independent components: water and energy, on which food depends. Water and energy in the modern world are highly independent (Khalkhali et al. 2018), even though they form a nexus in human dynamics. Proffering a solution to systems involving either, in the reality, results in a multi-objective problem. A good example in water resource management is the reservoir management system, where water competes for different purposes. In such cases, effectiveness of the optimization model for optimal allocation of water resources is of high importance. This is a noticeable problem inherent in both arid, semi-arid, and landlocked countries. Achieving optimal allocation of water resources to needed activities is hinged on real-time intelligent prediction of water level in the reservoir. Besides, obtaining long and short-term information about the status of water reservoirs require intelligent forecast.

Similarly, energy system modelling has taken different forms in the past few decades. Most models used can be classified based on their architecture into mathematical models and empirical models. The empirical models unravel pattern within a dataset to trigger informed decision-making as well as intelligent forecasts. Adaptive-neuro-fuzzy inference system (ANFIS) is one of the data-driven modelling techniques, which is a variation of the artificial neural network (ANN) developed to compensate for lack of subjective sensation inherent in ANN models. A global quest to proffer solutions to the energy trilemma (i.e. energy equity, energy security and environmental sustainability) has necessitated the use of alternative green energy sources, whose sustainability is hinged on real-time information about the state of the system both for operational and strategic decision-making. Achieving this in the era of the fourth industrial revolution requires intelligent predictive tools, of which ANFIS forms a part.

ANFIS model integrates ANN and fuzzy inference system (FIS) principles. It combines the strengths of both techniques on historical data with input-output fuzzification for the purpose of building a knowledge-based system. ANFIS has found its application in the past two decades in many fields: energy systems, medical diagnosis, econometrics, education, geology and so on. The integration between ANN and the Takagi-Sugeno-based FIS has also gained prominence in the field of water and energy resources. ANFIS has been applied in reservoir forecasting (Chang and Chang 2006; Hipni et al. 2013), groundwater level forecast (Sreekanth et al. 2011; Moosavi et al. 2014; Zare and Koch 2018) and water resource allocation (Abolpour et al. 2007) as well as in energy systems (Gayen and Jana 2017).

This chapter presents the relevance of ANFIS in energy and water resources with emphasis on methods of achieving an effective model using ANFIS and its hybrid. The relevance of time series data in energy and water resources is discussed in section 2. Section 3 focuses on black box models and their performance evaluation techniques. We also describe the basis of fuzzy inference system (section 4) as well as the structure of ANFIS model (section 5). Section 6 present speculations of the ANFIS model in water and energy resource modelling. Section 7 presents case studies of ordinary ANFIS model and hybrid-ANFIS model in water and energy resources, and section 8 concludes the chapter.

7.2 Time Series Data in Energy and Water Resources

Time series data is a chronologically ordered set of numerical observations with each observation associated with a time stamp. In water resource analysis, the application of time series as a statistical approach considered both surface and groundwater analyses. Through time series, hydrologic extremes such as floods and droughts can be analyzed. The approach of time series in water resource analysis entails the building-up of mathematical models to generate hydrological record, forecast events, detect trends, changes in hydrological profile, and find missing data. In similar manner, time series in energy system finds its relevance in design of energy profile, real-time forecast with online or off-line abilities, plant condition monitoring and so on. This statistical approach depends on historical data, which can factor the dynamic nature of the resources. However, the approach lacks empirical justification, and it ignores other factors such as socioeconomic activities that can impact on the availability of water and energy resource. A fundamental assumption on time series data follows that the series have a time invariant mean, which implies that it is homogeneous. Stationarity of the series; the absence of trends and shifts is another assumption. However, absolute stationarity of a time series data does not exist either in water or energy system modelling. Similar scenario exists in energy resource modelling. Other characterization which can be anticipated in water and energy system data are linearity and non-linearity, regularity and randomness, seasonality, saltation, chaotic property, fractality, cyclicity, and different degree of complexity (Tang et al. 2013). Energy systems consist of stochastic components which are non-deterministic. This is particularly a crucial component in renewable energy resources. This is because some renewable energy resource comprise climatological influential variables, which are more stochastic than deterministic. For example, changes in energy and water resources could occur gradually over a period, abruptly or take a more complex form. Though these changes are reflected in the data, however, the events leading to them may not be adequately factored into the empirical model of a time series. Chen and Boccelli (2018) concluded that a major limitation to time series data analysis is the identification of the appropriate mathematical structure (Chen and Boccelli 2018). To account for the impact of variables on the trend of time series data, models such as the autoregressive integrated moving average (ARIMA), multivariate time series (Tuncel and Baydogan 2018) and hybrid models such as combined autoregressive integrated moving average with explanatory variable (ARIMAX) combined with artificial neural network (Camelo et al. 2018) have been implemented. The auto-regressive models are dependent on their own history hence cannot provide reliable information about what never occurred in history or what is presently happening. The ARIMAX model weakens this limitation of the auto-regression but does not eliminate it. To adequately address some of these limitations associated with time series models in water and energy systems modelling, Azadeh et al. (2010) proposed an integrated fuzzy algorithm for regression analysis. Among the array of fuzzy algorithms is the ANFIS which is increasingly being adopted for modelling environmental time series data, particularly water resources. The rise in the use of ANFIS is due to the ease at which the model applies a set of fuzzy "if–then" rules with appropriate membership functions to generate a stipulated input-output mapping (Arora and Saini 2013), which is explained in section 5.

7.3 Black Box Models

Modelling of real systems has been colour-coded based on the explicitness of the model. Basically, there exist three categories, as described by Sjöberg et al. (1995). They include the black box, the grey box, and the white box models. While the black box models appropriately map inputs to output(s) from historical data for use in similar situations, the grey box models have a level of physical insight into the model with few parameter estimation from historical data (1995). On the other hand, the white box models are physical models with parameter estimation from mathematical equations describing the model.

Black box models, also called empirical models unravel hidden patterns and relationship existing within a set of observation, for example, in a time series data (with a high relevance in big data) Steindl and Pfeiffer (2017). They are completely data-driven without a need for prior knowledge of the system to be modelled. Unlike the white box models, black box models possess a high level of data abstraction, which enables them to perform optimally in non-linear systems. Examples of black box models include support vector machine (SVM), random forest (RF), multiple linear regression (MLR), ANN, ANFIS and so on. ANN, as an example of black box models, forms an integral part of the ANFIS model. Its self-learning and self-adaptive nature makes it suitable for use in establishing relationship between datasets (Bedi and Toshniwal 2019). It also consists of interconnected neurons with each neuron possessing a firing principle according to a rule. Three data learning techniques are commonly used in ANN models; the supervised learning (learning from a map of inputs to corresponding outputs), unsupervised learning (the use of an input to generate similar inputs in the same trend, often auto-regressive in nature), and the reinforced learning (learns and adapts to non-linear system through trial-and-error system from searching solution space as seen in the integration of evolutionary algorithms) (Kaelbling et al. 1996).

7.3.1 Performance Evaluation of Black Box Models

It is a common error to assume that a model which fits a historical dataset correctly will perform well on a new dataset. This anomality in black box modelling is often attributed to two common problems, which are underfitting and overfitting. Underfitting occurs when expected model parameters are missing, such that model over-describes the data than the information supported by the data (Van der Aalst et al. 2010). Overfitting is a situation where the model "memorizes" the data rather than learning the hidden pattern within. During overfitting there exists an excessive fine-tuning of the model to the data with more unnecessary parameters added to the model (Brown 2006). Overfitted models possess low variance with high bias unlike underfitting, which shows high variance and low bias. These violate the principle of Occam's razor, thereby assuming a correlation coefficient of unity (1) between the observed data and the model predicted data (Silva et al. 2018). When this occurs, the model performs poorly on new datasets. Black box modelling is often aimed at minimizing overfitting and underfitting, and maximizing model performance. This is often achieved through multiple trainings and re-training to ensure optimally adjusted network parameters without achieving the local minima. Indications for model overfitting include:

1. An approximately or exactly perfect correlation coefficient (i.e. $R = 1$) between training samples and network testing with same training samples
2. Low variance with high bias in results
3. Number of parameters in the model is the same as the total number of connections neurons and the bias terms (Hamzaçebi 2008).

Network overfitting problems are often solved using several approaches, some of which are briefly discussed as follows:

1. *Early Stopping*: This is one prominent means of preventing network overfitting in black box models. This technique tracks the model parameters during training process using a validation dataset – a selected sample data from the entire data. The training process is

stopped when the best obtained model performance over the validation dataset does not improve across a predefined training epoch (Ketkar 2017).

2. *Backwarding*: This technique is an improved form of early stopping. It involves network returning to the previous global optimal solution obtained within the solution space prior to overfitting. Best solution from the training and validation tests are kept during the network building process. However, as soon as overfitting starts occurring, the best-on-validation value is not updated when the best-on-training value is generated. This makes the network return to the best-on-validation value as soon as a stopping criterion is fulfilled (Silva et al. 2018). The network returns to the last generation before overfitting. This novel approach to semi-supervised learning was developed and applied to a Landsat 8 Operational Land Imager (OLI) image over Brazil, a location near the Amazon River by Silva et al. (2018). This was used to solve an image classification problem.
3. *Network Parameter to Sample Ratio*: This approach ensures that the training samples are more than the network parameters (Hiregoudar et al. 2011). However, learning of complex non-linear systems requires larger networks. In the case of a small amount of data with complex non-linearity, the use of Bayesian regularization becomes a viable option.
4. *Bayesian Regularization*: Regularization technique reduces the error from the new data by a systematic reduction in model capacity. It uses a portion of the model data kept as validation data for model control (Ketkar 2017). The Bayesian regularization technique has proven to be a potent tool in avoiding overfitting in ANN modelling (Das et al. 2010).
5. *Regularized Performance Function*: This incorporates the mean squared error (MSE), the squared network weight (MSW_i) and the network performance (μ) in the manner (Hamzaçebi 2008):

$$F = \mu \times MSE + (1 - \mu) \times MSW_i \tag{7.1}$$

This function applied to the training performance function results in small weights and biases as well as a smother network response. This in turn reduces the likelihood of network overfitting.

Performance metrics used to evaluate black box models depends on the purpose for which the model is used, either for classification or for data fitting. It also depends on the data type of the response variable, whether continuous or discrete. For data fitting problems with continuous response variables, statistical error estimates between the model forecast and the observed values of model variables are often used. Among the commonly used performance metrics are the following:

Root Mean Square Error (RMSE)

$$RMSE = \sqrt{\frac{\sum_{1=1}^{N}\left[y_i - \widehat{y_i}\right]^2}{N}} \tag{7.2}$$

Mean Absolute Percentage Error (MAPE)

$$MAPE = \frac{1}{N}\sum_{i=1}^{N}\left|\frac{y_i - \widehat{y_i}}{y_i}\right| \times 100\% \tag{7.3}$$

Mean Absolute Deviation (MAD)

$$MAD = \frac{1}{N}\sum_{i=1}^{N}\left|y_i - \mu\right| \tag{7.4}$$

Modified Nash-Sutcliffe Efficiency (MNSE)

$$MNSE = \left[1 - \frac{\sum_{i=1}^{N}\left|y_i - \widehat{y_i}\right|}{\sum_{i=1}^{N}\left|y_i - \overline{y}\right|}\right] \times 100\% \tag{7.5}$$

Determination Coefficient (R^2)

$$R^2 = \left[1 - \frac{variance\left(y_i - \widehat{y_i}\right)}{variance\left(y\right)}\right] \times 100\% \tag{7.6}$$

Relative Error (RE)

$$RE = \left[\frac{\left(y - \hat{y}\right)}{y}\right] \times 100\% \tag{7.7}$$

where

y = observed value

$\hat{y}$ = predicted value

y_i = each observed value for sample size $i = 1 \ldots N$

$\hat{y}_i$ = each predicted value for sample size $i = 1 \ldots N$

μ = mean of the observed values

Black-box models for classification problems with discrete response variable adopt a different method for model evaluation. Further reading on this technique can be found in Tharwat (2018).

7.4 Fuzzy Inference System

Many real-life decision-making problems involve linguistic variables whose measure of exactness could be controversial based on individual perceptions. Such problems use linguistic variables associated with imprecision. Fuzzy inference system (FIS) is a framework designed to handle such problems. It aims at representing a given system through the knowledge of the experts or past available data of the system to make sound inference from rule-based premises using the principle of fuzzy set theory. FIS model consists of three parts; the rule base, the database, and the reasoning mechanism. The inference engine of an FIS includes a set of fuzzy "if–then" rules. The first step in a FIS is the fuzzification of antecedent using fuzzy membership functions. In environmental system analysis, FIS has been applied extensively. Ponnambalam and Mousavi (2001) developed a FIS system for reservoirs operation based on optimal releases to maintain water quality in a creek and compared the result with a regression-based model. The FIS produced more reliable result than the regression model. Danso-Amoako and Prasad used FIS to predict the accumulation alkali metals in water (Danso-Amoako and Prasad 2015). The FIS model prediction was low, but when a hybrid FIS-genetic algorithm model was implemented, the model prediction accuracy increased. They concluded that hierarchical data-driven FIS model prediction accuracy was higher than the hierarchical rule-based expert system. As presented, the FIS model has some caveats. Ponnambalam and Mousavi (2001) concluded that FIS requires many parameters and a more sophisticated fitting method for parameter estimation. One way of reducing the need for sophisticated fitting method is the integration of neural network. The combination of ANN and FIS gives a model called the ANFIS. This has superior computation capability, better prediction accuracy and short computation duration compared to the FIS only (Atmaca et al. 2001). In-depth analysis of the ANFIS structure is presented in section 5.

7.5 Adaptive Neuro-fuzzy Inference System (ANFIS)

The ANFIS model is a multi-layer feedforward network with five layers: the fuzzy layer, the product layer, normalized layer, de-fuzzy layer and the total output layer, as shown in Figure 7.1 (Jang 1993; Rosadi et al. 2013). The ANFIS model applies the Takagi-Sugeno fuzzy system to map inputs to output space using a five-layer architecture. This technique uses hybrid learning rule comprising back-propagation gradient descent and least square methods for model premise and consequent parameter optimization.

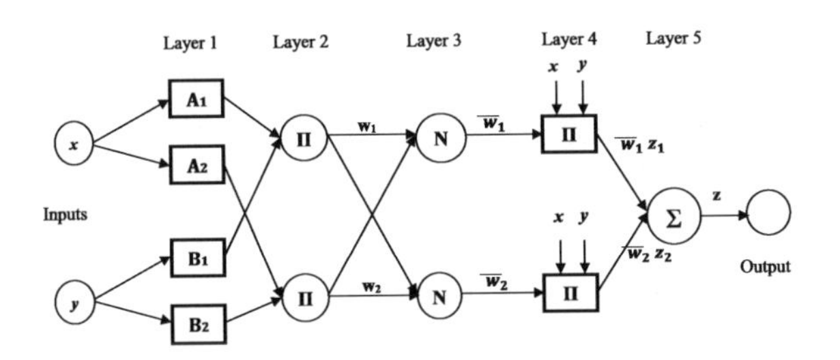

FIGURE 7.1
ANFIS model architecture

Layer 1 consists of fuzzy membership functions with output functions for each node, represented as:

$$O_i^1 = \mu_{A_i}(x),\ i = 1,2 \tag{7.8}$$

$$O_i^1 = \mu_{B_i}(y),\ i = 1,2 \tag{7.9}$$

Layer 2 computes the firing strength of a rule using a multiplicative operator as:

$$O_i^2 = w_i = \mu_{A_i}(x) \cdot \mu_{B_i}(y),\ i = 1,2 \tag{7.10}$$

Layer 3 normalizes the firing strength at the ith node of the structure using the ratio between the firing strength in the ith node and the sum of all firing strengths from all the rules (equation 7.11). Nodes in this layer are non-adaptive.

$$O_i^3 = \overline{w_i} = \frac{w_i}{w_1 + w_2} \qquad i = 1,2 \tag{7.11}$$

Layer 4 uses a nodal function to calculate the effect of the ith rule towards the output of the model using:

$$O_i^4 = \overline{w_i}\left(p_i x + q_i y + r_i\right) = \overline{w_i} z_i \tag{7.12}$$

Where p_i, q_i, r_i is a parameter set of the node and $\overline{w_i}$ is the normalized firing strength of the third layer.

Layer 5 has a single non-adaptive node, which calculates the overall output of the ANFIS model using a summation operation (Suparta and Alhasa 2016):

$$O_i^5 = \sum_i \overline{w_i} z_i = \frac{\sum_i w_i z_i}{\sum_i w_i} \tag{7.13}$$

Fundamental to ANFIS models is data clustering. Data clustering, groups similar patterns of data into clusters based on attributes and result of distance metrics (e.g. Mahalanobis, Euclidean, Minkowsky, Hellinger distances and so on) between each data cluster. Data clustering as a data mining technique occurs in two versions: the partitioning or hierarchical, approach: the fuzzy or crisp, mode: on-line or offline (Moertini 2002).

Many studies in the use of ANFIS for energy and water resources use the off-line mode for data clustering, associated with supervised learning. Data presented to the network learns repeatedly with the same training datasets until a lower network error is achieved without model overfitting (offline learning). However, an online training, involving model updating with recent data for retraining is highly necessary. This enables the network learn recent pattern and forecast not based on out-dated adjusted network parameters (Mamat et al. 2016). The possibility of network overfitting should therefore be carefully avoided to ensure a learning model is developed rather than a "memory bank."

7.6 Peculiarities of ANFIS Model

Every model has underlying assumptions. These assumptions largely determine the tractability and accuracy of the model. For data-driven models, assumptions may be for the data or the model itself. Researchers often assume that collected data fulfils these assumptions, which if false could make the model less of a microcosm of the larger system. ANFIS model is non-different from any other models in this regard. We consider some basic assumptions to be taken into consideration for use of ANFIS as a modelling technique:

1. There must exist set of numerical variables of model inputs and output.
2. ANFIS models follow the Takagi Sugeno-type fuzzy inference system, limiting the output to either a linear function or a constant.
3. ANFIS model works only on the zeroth or first-order Sugeno-type systems.
4. Only a single output is obtained from a weighted average defuzzification algorithm (wtaver).
5. No rule sharing (i.e. the number of output membership functions must equal the number of rules).
6. Unity weight must exist in each rule.

The efficiency and effectiveness of ANFIS model depend on optimal decisions in principal model components, which include the choice of the number and type membership function (Ortas 2013). However, it was discovered that the choice of clustering technique also plays a vital role most especially when model complexity is to be minimized and computational efficiency maximized. For example, grid partitioning technique for data clustering results in exponential increase in the rule base, which leads to what is called "curse of dimensionality" (Vasileva-Stojanovska et al. 2015). This clustering technique is also associated with the following features:

1. It generates complex models even for simple problems (Shihabudheen and Pillai 2018). This is seen in the number of rule database it generates, even for lower number of membership functions.
2. Due to the large rule base, it is computationally time-consuming and often results in the software used running out of memory and this may result into model in-effectiveness.

However, due to the versatility and robustness of ANFIS model in solving non-linear problems, modifications in the algorithm now exist in terms of its method of training as well as its learning time. An efficient and effective ANFIS model is expected to self-adjust with minimum global error and best global optimum, computationally efficient, have fast learning rate, and online adaptability (Shihabudheen and Pillai 2018). In order to achieve these, researchers have employed the following methods:

1. ANFIS model have been trained with an evolutionary algorithm (e.g. genetic algorithm, particle swarm optimization, differential evolution (Rezakazemi et al. 2017) for the purpose of parameter optimization and quick convergence.
2. The use of ANFIS hybridized with differential evolution algorithm to ensure global optimal solution without a local search with minimal tendency to converge prematurely.
3. Data pre-processing and decomposition into noise and signal using known algorithms (e.g. wavelet, singular spectrum analysis). This is to ensure the model learns principal patterns and not noise component for good generalization.

It is noteworthy that the ANFIS models are highly efficient in modelling complex non-linear problems, however selection of model parameters as well as clustering techniques predicates their efficiency and effectiveness.

7.7 Case Studies in Energy and Water Resource Modelling

With the strengths of ANFIS model as discussed in the previous section, researchers have both adopted the three modes of ANFIS training (Karaboga and Kaya 2018): the hybrid, derivative and heuristic-based training techniques in energy and water resource management studies. This section examines some cases using examples in energy and water resource management.

7.7.1 ANFIS Models in Energy and Water System

Water level in reservoirs can be predicted based on historical data using ANFIS model. Its use was investigated by Chang and Chang (2006), where reservoir water level was forecast during flooding situations. An ANFIS structure with 132 typhoon events for a period of 31 years was collected. This amount to one of the largest datasets for ANFIS model in the literature, comprising 8,640 elements available for training the model. The dataset was divided into training, verification, and testing data comprising of 62, 38 and 32 events, respectively. Three hours shifted time delay inputs were used as network inputs with the flow of the river as response. With repeated trainings for weight adjustment, two models were explored; one with upstream flow patterns and its corresponding outflow as input variables and the other without reservoir outflow pattern as model inputs. Subtractive fuzzy clustering was applied, which significantly reduced the rule base. The first model with a time delay of one yielded four rules with the lowest mean absolute error.

Similar to the work of Chang and Chang (2006), Anusree and Varghese (2016) explored three models – adaptive neuro-fuzzy inference system (ANFIS), artificial neural network (ANN) and multiple non-linear regression (MNLR) – to predict streamflow in the Karuvannur River Basin. The study used precipitations and flow data obtained from nine rain gauge stations with different time lags. The ANFIS model with its lowest root mean square error recorded has the highest Nash-Sutcliffe model efficiency values for all the five cases considered (0.0384, 0.9029, 0.8683, 0.9028, 0.9536, respectively) compared to the ANN (−0.8344, 0.8474, 0.9057, 0.9096, 0.9066, respectively) and MNLR (0.5519, 0.9218, 0.9176, 0.9221, 0.9344, respectively) models.

ANFIS has also been used in a couple of studies in energy system for long and short-term forecast. One of the recent studies in this domain is application of ANFIS for a multi-stage forecast for a wind farm Zheng et al. (2017). The study integrated meteorological data and data obtained from the wind farm Supervisory Control and Data Acquisition (SCADA) system to forecast short-term wind power. The first stage tunes ANFIS parameters using Particle Swarm Optimization (PSO) algorithm to effectively map meteorological variables like wind speed, wind direction, air pressure, air temperature and humidity to the actual wind speed obtained from the SCADA system. The second stage of the model uses the trained model for short term forecast. In the study, ANFIS was compared with Double-Staged Neural Network (DSN), Double Stage Hybrid Genetic Algorithm-Neural Network (DSHGANN) and Double Stage Hybrid PSO-Neural Network (DSHPSONN). ANFIS outperformed the other three algorithms with which it was compared with a Mean Absolute Percentage Error (MAPE) of 8.1133 %.

The ANFIS model has also been used at end-user level to model lighting load profile in residential houses Popoola (2016). Socio-demographic factors like the active occupancy and income were integrated with the ambient lighting level as inputs and light usage as output. The result obtained was compared to a regression model for same situation. The ANFIS model outperformed the regression model by an R^2 value between 1.7 % and 24.6 %.

7.7.2 Hybrid ANFIS in Energy and Water Resource Modelling

ANFIS modelling within the past decade has been hybridized using two or all the three common training techniques for consequent parameter optimization in energy and water resource modelling. These include the use of evolutionary algorithms, wavelet, machine learning functions and so on. Further reading on this can be found in Shihabudheen and Pillai (2018). The hybridization of ANFIS models is an error minimization technique to obtain Pareto optimal parameters for membership function. Examples of the applications of these two are considered below.

Water allocation from a river basin in Iran was performed using a ANFIS hybridized with adaptive reinforced learning by Abolpour et al. (2007). The study used three inputs: surface water, water demand and groundwater from each sub-basin a reinforced learning algorithm, which tunes the ANFSI model parameters in the second and fourth layer. This model predicts different river flow in the seven sub-basins considered. The adaptive neuro-fuzzy reinforcement learning (ANFRL) technique optimizes a model for each sub-basin whose optimum decision variables have been pre-obtained. The details of the algorithm can be obtained from Abolpour et al. (2007).

Another hybrid model incorporating evolutionary algorithm and differential evolution, comparing the result with ordinary ANFIS model was presented by Azad et al. (2017). This study integrated ant colony for continuous domain optimization technique (ACO) and differential evolutionary algorithm with ANFIS model for parameter tuning of the first and fourth layer. The resulting model was used to forecast water quality of Gorganrood River in Golestan Province, Northern Iran. The model determines the best water quality parameters suitable for the ANFIS model, same for the other two models using sensitivity analysis. Seven groups with constant river flow and variables (HCO_3, Ca, Cl, SO_4, Mg, Na, and K) to predict quality parameters (electrical conductivity (EC), total hardness (TH) and sodium absorption ratio (SAR)). Different models performed optimally for different water quality measures. While the ANFIS-GA and ANFIS-DE showed the best performance for SAR, than ANFIS and ANFIS-ACO. The MAPE values of 5.16 and 9.55 were obtained for EC and TH quality parameters respectively.

Time series as earlier discussed are associated with noise and signal component. A balance between quality and quantity of data is highly essential before samples can be a microcosm of the system. The use of ANFIS model integrated with wavelet-based denoising approach has also gained significance in water resources management. The study (Nourani and Partoviyan 2018) is a quintessential example of this. The study uses a wavelet-based denoising technique to smooth hydrological data comprising daily and multi-step ahead rainfall run-offs of two stations: Milledgeville and Pole Saheb located in Oconee River, USA and Jighatu River watershed, Iran, respectively.

Similar to the work of Nourani and Partoviyan (2018) is the application of wavelet transform integrated with ANN and ANFIS to model ground level fluctuations by Zare and Koch (2018). With all indications of non-stationarity in the data, wavelet transform integrated with ANFIS enhanced the model result. With the use of fuzzy c-means clustering technique, the number of rules generated was minimal, thereby reducing model complexity. The denoising and jittering resulted in performance improvement of ANN and ANFIS by 14 % and 12 % for Milledgeville Station and by 22 % and 16 % for Pole Saheb Station.

Hybrid ANFIS model in energy systems also follows similar trends. ANFIS have been hybridized with PSO, genetic algorithm (GA) and differential evolution (DE) in predicting monthly solar radiation using sunshine radiation, maximum and minimum air temperature, monthly rainfall and clearness index as inputs Halabi et al. (2018). The ANFIS-PSO model outperformed all other models with an R^2 value of 0.2482 and 1.4097 MAPE value.

Training ANFIS with hybrid techniques improves parameters in the adaptive layers towards achieving a global minimum error, quick convergence, and effectiveness at a lower throughput.

7.8 Conclusion

ANFIS models can best be considered when data-driven decision-making is to be made, with data exhibiting a high level of uncertainty. The success of ANFIS model in terms of computational simplicity and early convergence is dependent on the type of clustering technique used. Time series data possesses two components, the signal, and the noise. It is highly essential that the data for ANFIS modelling be pre-processed such that noise component is eliminated or its effect reduced to the minimum without loss of data quality. Data re-construction techniques like singular spectrum analysis (SSA) and the wavelet-based approach are potent tools for this. Parameter optimization is highly essential in ANFIS modelling to enhance network generalization. Hybrid ANFIS models with online training is recommended for energy and water resource modelling engineers for robust model performance in highly complex non-linear systems.

References

Abolpour, B., Javan, M., & Karamouz, M. (2007). Water allocation improvement in river basin using adaptive neural fuzzy reinforcement learning approach. *Appl. Soft Comput. J., 7*(1), 265–285.

Anusree, K., & Varghese, K.O. (2016). Streamflow prediction of Karuvannur river basin using ANFIS, ANN and MNLR Models. *Procedia Technol., 24*, 101–108.

Arora, N., & Saini, J.R. (2013). Time series model for bankruptcy prediction via adaptive neuro-fuzzy inference system. *Int. J. hybrid Inf. Technol., 6*, 14.

Atmaca, H., Catisli, B., & Yavuz, H.S. (2001). The comparison of fuzzy inference systems and neural network approaches with ANFIS method for fuel consumption data. *Second International Conference on Electrical and Electronics Engineering Papers ELECO 2001*, Bursa, Turkey.

Azad, A., Karami, H., Farzin, S., Saeedian, A., Kashi, H., & Sayyahi, F. (2017). Prediction of water quality parameters using ANFIS optimized by intelligence algorithms (Case study: Gorganrood River). *KSCE J. Civ. Eng.*, 1–8.

Azadeh, A., Saberi, M., & Seray, O. (2010). An integrated fuzzy regression algorithm for energy consumption estimation with non-stationary data: A case study of Iran. *Energy, 35*(11).

Bedi, J., & Toshniwal, D. (2019). Deep learning framework to forecast electricity demand. *Appl. Energy, 238* (October), 1312–1326.

Brown, T.A. (2006). *Confirmatory factor analysis for applied research.* Spring Street, New York: Guilford Publications.

Cai, X., Wallington, K., Jood, Shafiee, M., & Marston, L. (2018). Understanding and managing the food-energy-water nexus—opportunities for water resources research. *Adv. Water Resour., 111* (November), 259–273.

Camelo, H.D.N., Lucio, P.S., Leal Junior, J.B.V., Carvalho, P.C.M.D., & Santos, D.V.G.D. (2018). Innovative hybrid models for forecasting time series applied in wind generation based on the combination of time series models with artificial neural networks. *Energy, 151*, 347–357.

Chang, F.J., & Chang, Y.T. (2006). Adaptive neuro-fuzzy inference system for prediction of water level in reservoir. *Adv. Water Resour., 29*(1), 1–10.

Chen, J., & Boccelli, D.L. (2018). Real-time forecasting and visualization toolkit for multi-seasonal time series. *Environ. Model. Softw., 105*, 244–256.

Danso-Amoako, E., & Prasad, T.D. (2015). Using fuzzy inference system to predict Iron and manganese accumulation potential in water distribution networks. *Procedia Eng., 119*, 379–388.

Das, A., Maiti, J., & Banerjee, R.N. (2010). Process control strategies for a steel making furnace using ANN with Bayesian regularization and ANFIS. *Expert Syst. Appl., 37*(2), 1075–1085.

Gayen, P.K., & Jana, A. (2017). An ANFIS based improved control action for single phase utility or micro-grid connected battery energy storage system. *J. Clean. Prod., 164*, 1034–1049.

Halabi, L.M., Mekhilef, S., & Hossain, M. (2018). Performance evaluation of hybrid adaptive neuro-fuzzy inference system models for predicting monthly global solar radiation. *Appl. Energy, 213* (January), 247–261.

Hamzaçebi, C. (2008). Improving artificial neural networks' performance in seasonal time series forecasting. *Inf. Sci. (Ny).*, *178*(23), 4550–4559.

Hipni, A., El-shafie, A., Najah, A., Karim, O. A., Hussain, A., & Mukhlisin, M. (2013). Daily forecasting of dam water levels: Comparing a Support Vector Machine (SVM) model with Adaptive Neuro Fuzzy Inference System (ANFIS). *Water Resour. Manag.*, *27*(10), 3803–3823.

Hiregoudar, S., Udhaykumar, R., Ramappa, K. T., Shreshta, B., Meda, V., & Anantachar, M. (2011). Artificial neural network for assessment of grain losses for paddy combine harvester a novel approach. *Commun. Comput. Inf. Sci.* (CCIS), *140*, 221–231.

Jang, J.S.R. (1993). ANFIS: Adaptive-network-based fuzzy inference system. *IEEE Trans. Syst. Man. Cybern.*, *23*(3), 665–685.

Kaelbling, L. P., Littman, M. L., & Moore, A. W. (1996). Reinforcement learning: A survey. *J. Artif. Intell. Res.*, *4*, 237–285.

Karaboga, D., & Kaya, E. (2018). Adaptive network based fuzzy inference system (ANFIS) training approaches: A comprehensive survey. *Artif. Intell. Rev.*, 1–31.

Ketkar, N. (2017). Regularization techniques. In *Deep learning with python: A hands-on introduction*, San Francisco, CA: Apress Media, pp. 209–214.

Khalkhali, M., Westphal, K., & Mo, W. (2018). The water-energy nexus at water supply and its implications on the integrated water and energy management. *Sci. Total Environ.*, *636*, 1257–1267.

Mamat, M., Porle, R. R., Parimon, N., & Islam, N. M. (2016). An adaptive learning radial basis function neural network for online time series forecasting. In *Advances in machine learning and signal processing*, P. J. Soh et al., ed. Switzerland: Springer International Publishing Switzerland, pp. 25–34.

Moertini, V.S. (2002). Introduction to five data clustering algorithms. *Integral*, *7*(2), 87–96.

Moosavi, V., Vafakhah, M., Shirmohammadi, B., & Ranjbar, M. (2014). Optimization of wavelet-ANFIS and wavelet-ANN hybrid models by Taguchi method for groundwater level forecasting. *Arab. J. Sci. Eng.*, *39*(3), 1785–1796.

Nourani, V., & Partoviyan, A. (2018). Hybrid denoising-jittering data pre-processing approach to enhance multi-step-ahead rainfall–runoff modeling. *Stoch. Environ. Res. Risk Assess.*, *32*(2), 545–562.

Ortas, I. (2013). Influences of nitrogen and potassium fertilizer rates on pepper and tomato yield and nutrient uptake under field conditions. *Acad. Journals*, *8*(23), 1048–1055.

Ponnambalam, K., & Mousavi, S. J. (2001). Water quality management using a fuzzy inference system. *J. water Manag. Model.*, 10.

Popoola, O. M. (2016). Modeling of residential lighting load profile using adaptive neuro fuzzy inference system (ANFIS). *Int. J. Green Energy*, *13*(14), 1473–1482.

Rezakazemi, M., Dashti, A., Asghari, M., & Shirazian, S. (2017). H2-selective mixed matrix membranes modeling using ANFIS, PSO-ANFIS, GA-ANFIS. *Int. J. Hydrogen Energy*, *42*(22), 15211–15225.

Rosadi, D., Subanar, T., & Suhartono. (2013). Analysis of financial time series data using Adaptive Neuro Fuzzy Inference System (ANFIS). *Int. J. Comput. Sci. Issues*, *10*(2), 491–496.

Shihabudheen, K. V and Pillai, G.N. (2018). Knowledge-based systems recent advances in neuro-fuzzy system: A survey, *Knowledge-Based Syst.*, 1–27.

Silva, S., Vanneschi, L.A.I., Cabral, R., & Vasconcelos, M.J. (2018). A semi-supervised genetic programming method for dealing with noisy labels and hidden overfitting. *Swarm Evol. Comput., 39* (September), 323–338.

Sjöberg, J., et al. (1995). Nonlinear black-box modeling in system identification: A unified overview. *Automatica, 31*(12), 1691–1724.

Smajgl, A., Ward, J., & Pluschke, L. (2016). The water-food-energy nexus—realising a new paradigm. *J. Hydrol., 533*, 533–540.

Sreekanth, P.D., Sreedevi, P.D., Ahmed, S., & Geethanjali, N. (2011). Comparison of FFNN and ANFIS models for estimating groundwater level. *Environ. Earth Sci., 62*(6), 1301–1310.

Steindl, G., & Pfeiffer, C. (2017). Comparison of black box models for load profile generation of district heating networks, in SDEWES- 12th Conference on Sustainable Development of Energy, Water and Environment Systems, pp. 1–10.

Suparta, W., & Alhasa, K.M. (2016). Adaptive neuro-fuzzy inference system. *Modeling of Tropospheric Delays Using ANFIS* (2009), 5–19.

Tang, L., Wang, C., & Wang, S. (2013). Energy time series data analysis based on a novel integrated data characteristics testing approach, *Procedia Comput. Sci.*, 759–769.

Tharwat, A. (2018). Classification assessment methods. *Appl. Comput. Informatics*, Article in Press.

Tuncel, K.S., & Baydogan, M.G. (2018). Autoregressive forests for multivariate time series modeling. *Pattern Recognit., 73*, 202–215.

van der Aalst, W.M.P., Rubin, V.H.M., Verbeek, W., van Dongen, B.F., Kindler, E., & Gunther, C.W. (2010). Process mining: A two-step approach to balance between underfitting and overfitting. *Softw Syst Model, 9*, 87–111.

Vasileva-Stojanovska, T., Vasileva, M., Malinovski, T., & Trajkovik, V. (2015). An ANFIS model of quality of experience prediction in education. *Appl. Soft Comput. J., 34*, 129–138.

Zare, M., & Koch, M. (2018). Groundwater level fluctuations simulation and prediction by ANFIS- and hybrid Wavelet-ANFIS/Fuzzy C-Means (FCM) clustering models: Application to the Miandarband plain, *J. Hydro-Environment Res., 18* (December), 63–76.

Zheng, D., Eseye, A.T., Zhang, J., & Li, H. (2017). Short-term wind power forecasting using a double-stage hierarchical ANFIS approach for energy management in microgrids. *Prot. Control Mod. Power Syst.*, 2(1), 13.

Index

A

Adaptive Neuro-Fuzzy Inference System 117, 128, 131, 133
Advanced Manufacturing 33, 34, 43
AGV 23–29
Alumina 91, 99
Aluminium 64, 80, 99
ANFIS 117–130
ANN 118, 120, 112, 124, 128, 130
ANOVA 76–78
Ant Colony Optimization 23–25, 27, 29
Artificial Neural Network 118, 119, 128
Automated Guided Vehicle 23, 25, 31
Automatic Material Handling 23

B

B4C 57–59, 64–67, 76–80
Bearing Parameter Identification 3, 10, 20
Bird Swarm Algorithm 83–85, 89, 92, 98, 100

C

C# 23, 26–29
CNT-Polymer Nanocomposite 13–16, 19, 20
Complex Non-Linear Systems 122, 130
Computational Time 9, 85, 95
Convergence Rate 85
Cuckoo Search Algorithm 8, 17

D

Data Envelopment Analysis 33, 35, 46, 52–55, 67
DEA 33–55, 67
Demand Side Management 104, 112, 115, 116
DSM 104, 105, 112, 114, 115

E

Effectiveness 34, 39, 48, 51, 85, 99, 105, 117, 126, 127, 130
Elastic Modulus 13–15, 17, 19
ELECTRE 57–59, 62, 65–68, 71, 72, 76, 79
Electric Discharge Machine 57, 60
Electricity Consumption 104–105, 115
Energy 16, 33–35, 47–55, 59–61, 80, 84, 96–97, 99, 103–110
Energy Resources 118–119

F

Firefly Optimization 3, 9, 19
FIS 118, 123–124
Fuzzy Inference System 117–118, 123, 131

H

HAZ 84, 91–98
Hole Taper 91–98

L

Laser Beam 84

M

Manufacturing Sector 33–35, 37–43, 46–57
Material Modelling 3, 4, 13, 19
Material Removal Rate 57–58, 66, 81
Metaheuristic Algorithm 4, 6, 20, 89
Metaheuristic Optimization 3, 19, 84
Metal Matrix Composite 80–81
Micro-Drilling 83–85, 91, 95, 98, 99
Mobile Robot 23–24
Multi-Objective Optimization 83, 95–96, 99

N

Nano-composite 3, 13–14, 16, 19–20
Novel Bat Algorithm 83–87, 92, 99

P

Parameter Selection 4
Performance Measurement 33, 35, 37–38, 40, 42, 45, 53
Process Parameters 19, 57, 58, 60, 64–65, 80, 83–85, 96

R

Radial Over-Cut 58, 77, 79
Randomness 23–24, 119
Reinforcement 13, 58, 64, 66, 76–78, 129–130, 132
Reliability 104
RSM 64, 85–86, 91–96, 99

S

Signal to Noise Ratio 76–77
Simulation 23–25, 31, 37, 44–46, 105, 112
Single Objective Optimization 83, 92–94
Social Behaviour 4, 85, 89
Stir casting 64
Surface Roughness 57–58, 65, 67, 76, 92, 93
Sustainable Manufacturing 35, 49–50
Swarm Intelligence 4, 109

T

Taguchi 58–59, 63, 79, 132
Thermal Conductivity 13
TiC 57–59, 63–67, 76–80
Time Series Data 117–120, 130
Tool Wear Rate 57–58, 65–66, 77
Transportation 23–24, 36, 47, 116

V

VABC 105, 110, 112
Velocity based Artificial Bee Colony Algorithm 105, 110, 112

W

Water Resources 117–119, 125, 129, 131